Liaison Functions 2019

Held at the 2019 AIChE Spring Meeting and
15th Global Congress on Process Safety

New Orleans, Louisiana, USA
31 March - 4 April 2019

ISBN: 978-1-5108-8697-1

Printed from e-media with permission by:

Curran Associates, Inc.
57 Morehouse Lane
Red Hook, NY 12571

Some format issues inherent in the e-media version may also appear in this print version.

Printed by Curran Associates, Inc. (2019)

AIChE
120 Wall Street, FL 23
New York, NY 10005-4020

Phone: (800) 242-4363
Fax: (203) 775-5177

www.aiche.org

Additional copies of this publication are available from:

Curran Associates, Inc.
57 Morehouse Lane
Red Hook, NY 12571 USA
Phone: 845-758-0400
Fax: 845-758-2633
Email: curran@proceedings.com
Web: www.proceedings.com

TABLE OF CONTENTS

547777 Using Process Safety Tools to Find and Sell an Inherently Safer Design: A Case Study

Using Process Safety Tools to find and sell an Inherently Safer Design: A Case Study

By

Claire Fluegeman and Margery Kosch

3M Company

This paper will focus on the use of PSM tools to communicate risk and influence decision making during the design process.

As our facility began the design and installation of a new large-scale process, we identified numerous challenges. A significant processing culture shift came when the new continuous operation was introduced into a production structure that is largely used to working with batch process operations. Included in this was a shift from using drumstock Acrylic Acid to using bulk tank storage and delivery to the process equipment. Receiving bulk Acrylic Acid in tank trucks and railcars presented new challenges, and new opportunities for error, that had not yet been considered.

During the detailed design process hazard analysis that was conducted for the project, a potentially catastrophic scenario was identified involving the unloading of Acrylic Acid. Because of the potential impact of the event, a Layers of Protection Analysis (LOPA) was conducted. Ultimately, the only true way to reduce the risk of the event to an acceptable level was to evaluate numerous modifications to the unloading process. However, this was met with a fair amount of resistance from the design team as it could significantly impact the project cost and implementation timetable.

The site PSM Coordinator, who facilitated both the PHA and the LOPAs for the project, had the responsibility to persist with those that make the decisions and ensure that the true risk of the scenario was properly communicated. Using the qualitative and semi-quantitative results of the PHA and LOPA, site safety personnel were able to intervene and convince the decision makers that the right thing to do was develop an inherently safer design for the unloading of this highly hazardous material. Even though there were additional costs and some redesign was involved, the final product resulted in an inherently safer design for unloading and delivering bulk acrylic acid, and the repurposing of a facility that was currently not in service.

548217 Should Chemical Engineers Have More Influence on Safety Than They Do Now?

This presentation will focus on the question of whether chemical engineers and other health, safety, and environmental professionals within the corporation possess enough influence and resources to effectively prevent major process safety hazards and will discuss what increases in authority may be appropriate.

548395 Courageous Process Safety Leader – a Bad Guy

An honest competent process safety engineer is considered to be the trouble maker and a job stopper. This is one of the most thankless role in the high hazard industries. If this role is performed honestly without taking into consideration the impact on the relationship with the management and employees, then the results are certainly positive and encouraging for a long run.

The author will share three real life examples with challenges and success stories. The first example is from a corporate world, he influenced senior management about the benefits of process safety, where none of their operations were covered by OSHA-PSM. The second example is from the major projects, an engineering and design world, where he influenced project directors, projects managers and design engineers to implement inherent safety in design in projects. The last example is from an operating refinery, where he changed the process safety culture at the management as well as shop floor level by implementing effectively a work permit system which was completely broken. The author faced several challenges and threats which impacted his personal life in addition to the professional relationships.

This paper will be concluded by sharing the Do's and Don'ts for transforming the process safety culture from corporate world to engineering and design and down to the shop floor in complex high hazard industries.

552663 Inclusion & Diversity: An Innovation Imperative

Fostering an inclusive work environment is a core business value at Chevron. Our strong belief is that an organization must welcome diversity of thought, life experiences and work styles to succeed – and an innovative esprit de corps is imperative to build on that success. Chevron Oronite President Bruce Chinn talks about the foundational role of inclusion & diversity in the innovation imperative.

550560 Industrial Distillation - Theory to Practice

This paper deals with a short history, and the basics of distillation as a Unit Operation. Distillation theory is converted into practice and how to optimally simulate and choose the appropriate number of stages for a tower.

550570 Dividing Wall Columns - a Primer

This paper presents an introduction to dividing wall columns and their applications in the chemical process industry

550899 Comparing Membrane & Distillation on an Equivalent Basis

Comparing Membrane & Distillation on an Equivalent Basis
Adrian Chavez Velasco, Rakesh Agrawal
Davidson School of Chemical Engineering, Purdue University, West Lafayette, IN

Among the separation technologies portfolio, distillation continues to be the preferable method for majority of the separations in the Chemical and Petrochemical plants. Just in the U.S. alone, it is estimated that about 90-95 % of all separations in chemical plants and refineries are performed by means of distillation columns [1]. Although distillation is widely used, it is frequently perceived to have a low energy efficiency, whereas in contrast, membrane separations are classified as a highly energy efficient processes which could potentially replace distillations on a wide scale [2], [3] [4].

However, this conclusion is not well supported since it has been mainly derived from comparisons made only with respect to conventional distillation, ignoring other operation modes that could be more energy efficient. Apparently, there is a widespread confusion in considering distillation as an absolute heat driven device, which sometimes requires a high amount of thermal energy input at the reboiler. The comparison between work driven membranes and heat driven distillations often leads to inaccurate conclusions. In reality, it is feasible to operate distillation solely with work input without requiring any external heat [5]. Our aim is to compare solely work driven distillations with membranes to avoid comparing two different forms of energy with each other.

The current research work centers on the development of a consistent procedure to compare the energy performance between membranes and distillation. This methodology comprises the use of rigorous simulation, and the optimization of both processes as an attempt to evaluate them at their best operating conditions. The proposed method was applied to the analysis of two important separation cases; p-xylene/o-xylene and propane/propylene. The results showed that the use of membranes with demonstrated permselectivity values does not always yield to the highest energy efficiency. Indeed, for the studied mixtures, the operation of distillation with work input consumes about an order of magnitude less energy than membranes for the case of p-x/o-xylene and ~27% less energy for the separation of propane/propylene under the conditions encountered at industrial scale (high recovery and purity). Membranes certainly offer some advantages over distillation in terms of energy efficiency, but for the analyzed cases, it only happens when the recovery is moderate or when the inlet is quite enriched in the most permeable component.

The extension of the developed methodology for the separation of other mixtures would enable researchers and practitioners to correctly identify under which conditions the use of either membranes or distillation would bring into the highest energy savings among them, thus contributing to design and operation of efficient separation processes.

Bibliography

[1] U.S. Department of Energy, "Materials Research for Separation Technologies: Energy and Emission Reduction Opportunities," 2005.
[2] D. S. Sholl and R. P. Lively, "Seven chemical separations to change the world," Nature, vol. 532, pp. 435-437, 2016.
[3] J. W. Koros, "Evolving Beyond the Thermal Age of Separation Processes: Membranes Can Lead the Way," AIChE Journal, vol. 50, no. 10, pp. 2326-2334, 2004.
[4] R. P. Lively and D. S. Sholl, "From water to organics in membrane separations," Nature Materials, vol. 16, pp. 276-279, 2017.
[5] A. A. Shenvi, D. M. Herron and R. Agrawal, "Energy Efficiency Limitations of the Conventional Heat Integrated Distillation Column (HIDiC) Configuration for Binary Distillation," Ind. Eng. Chem. Res., vol. 50, pp. 119-130, 2011.

550597 Column Inspections and Safety

Distillation column inspections are an essential step to a successful installation or revamp. Safety, mechanical design and knowledge of the process intended for the column are just a few of the topics that must be covered before entering the column. Setting up a checklist of items to be checked alone often requires significant attention to detail. This presentation will provide an introduction to preparation for a safe and successful column inspection.

550621 Distillation Tower Troubleshooting Tips for Beginners

Have you been called upon to troubleshoot a problem tower, but do not know where to begin?

This presentation is aimed at providing beginner tips and techniques for troubleshooting distillation columns.

550564 Radioisotope Techniques for Process Investigations

In all chemical plants and refineries there generally is not enough instrumentation in the right places in our processes. During the design phase, decisions were made between knowing more about the process, which means buying more instruments, and saving capital funds. When a problem appears with the operation of a vessel, there are always gaps in the knowledge of what is happening due to a lack of instruments in the places where knowledge is not routinely required. Process engineers and operations personnel need to know what is happening so they can take corrective action. When knowledge is missing, they often take actions that make matters worse.

For distillation and separation processes Chemical Engineers have historically relied upon plant process measurements such as flow rates, temperatures, and pressures and model or simulation results for data with which to solve troubleshooting or re-design problems. An array of on-line diagnostic services are available that can provide additional data offering real-time information on how pieces of process equipment are actually operating. The most common application is Gamma Scanning. This test is primarily applied, but not necessarily limited to, distillation or separation columns. Gamma scans provide a density profile of the internal process of operating distillation columns and other process vessels. The density profile can be used to diagnose the hydraulic operating conditions of mass transfer devices such as damage to internals, flooding, degree of entrainment or weeping, liquid levels on trays and distributors, liquid distribution through packed beds, etc. The presenter will show case studies where scanning revealed vital process information that helped solve an operating problem, or helped make a revamp/re-design successful. Additional derivatives from gamma scanning of detecting flow distribution patterns will be demonstrated. The ThruVision Scan technique is a specialized horizontal gamma scan used to generate a topographic profile of the internal cross-sectional density of process equipment. This profile is useful for the detailed study of liquid flow distribution through packed columns as well as having applications beyond distillation columns. A hybrid application involving tracers and scanning on fixed bed catalytic reactors will also be discussed.

543485 Remediation of Cr (VI)-Contaminated Soil Using Hydrazine

Abstract

The current status of technology for remediation of soil contaminated with Cr (VI) and our work on decontamination using hydrazine as the reducing agent will be described. Hexavalent chromium is a major metal-derived pollutant of water and soil which is toxic and established carcinogen. Mining and metallurgical industries activities as well as chrome chemicals and tanneries are major sources of release to the environment. Sites contaminated with Cr (VI) exist in countries such as USA, South Africa, Kazakhstan, New Zealand, UK, Canada, Italy and India.

Remediation of Cr (VI) essentially requires a suitable reducing agent or technique that converts it to much less toxic Cr(III). Both chemical reduction and bio-remediation have been practiced. The reducing agent should be non-toxic and any excess used should easily decompose. Use of calcium polysulfide, ferrous sulfate, sodium bisulfite, sodium dithionite, zero-valent iron and organic reducing agents such as molasses and glucose have been reported.

We have used hydrazine, a novel reducing agent, for remediation of Cr (VI) in soil. The reduction reaction is very fast and any excess can be easily taken care of. Remediation may be done in-situ by injecting the reagent at the site and ensuring its distribution in the soil mass. Alternatively, it may be done ex-situ either by mixing the soil for a sufficient time or by washing the Cr (VI) out of the soil and treating it with hydrazine in a stirred vessel or reactor. The last option is most advisable since the soil can be made free from both forms of chromium, Cr (VI) and Cr (III).

A literature report on the reaction of Cr (VI) with hydrazine is scanty. Kinetics of the reaction which is essential for comparison of its performance with competing reagents as well as for design of a reactor have not been reported so far. In this work we have done this study for remediation of soil washings. Soil slurries with different initial Cr (VI) concentrations were treated with hydrazine hydrate in a stirred reactor at different mole ratios. Samples were withdrawn from time to time, the suspended solids filtered out and the unreacted Cr (VI) estimated using the diphenylcarbazide technique. Reaction rates for different initial concentrations of Cr (VI), mole ratio of the reducing agents, pH and temperature were measured. The reaction was found to follow second order kinetics. The rate equation was developed by fitting the experimental data obtained for varying the initial concentrations of Cr(VI) and hydrazine at a given temperature and pH . The data fit was very good with a correlation coefficient of 0.984. The activation energy was determined from the Arrhenius plot. Further the second order rate constant was found to be linear in pH of the reaction medium.

550530 Building Your Career Army

When it comes to career advancement and planning you often hear about networking. Most people think of networking as making contents, attending networking events or grabbing one lunch with someone. Most people also think networking only matters when you are looking for a job. Networking, however is more strategic and more involved than just that. It's about building an army of people in your career ecosystem that can help you strategize, execute, and fight for your career day to day. In this session, we will talk about what a career army looks like, who you need to enlist, and how to mobilize your troops to help you advance your career.

550559 Mentoring Vs Coaching, What Is the Difference? What Is the Benefit of Each?

Everyone thinks they need or want a mentor. You actually may need a coach or both a coach and mentor. You may think there is little difference between coaching and mentoring. You may think they are similar or that they are the same. They are different and we will go over those differences. When do you need one vs the other. We will also go over the tips for successful interactions in a mentor, mentee relationship. Roles and responsibilities of the mentor & mentee will be looked at. Plus a personal view from the trenches of having both successful and non successful coaches, mentors, and mentees.

552664 Innovation and Emerging Technology Perspectives in Hydrogen and Fuel Cells at the U.S. Department of Energy

Today the technology around generating efficient and sustainable energy is rapidly evolving and hydrogen and fuel cells are emerging, versatile examples within a portfolio of options. Dr. Sunita Satyapal will provide an overview of the U.S. Department of Energy's (DOE) activities within the Office of Energy Efficiency and Renewable Energy (EERE) focusing on hydrogen and fuel cell technologies. This is an exciting time in the fuel cell industry with several thousand commercial fuel cell vehicles now sold or leased, hundreds of fueling stations worldwide, and more than a quarter of a million stationary fuel cells providing clean, reliable power. DOE's Fuel Cell Technologies Office (FCTO) addresses key technical barriers faced by hydrogen and fuel cells through a comprehensive portfolio of early-stage research and development (R&D) with the potential to meet technical and cost targets that enable competitiveness with incumbent technologies in the market.

The presentation will cover DOE's H2@Scale initiative which will enable innovations to generate cost-competitive hydrogen as an energy carrier, coupling renewables, as well as nuclear, fossil fuels, and the grid, to enhance the economics of both baseload power plants and intermittent solar and wind, enhancing resiliency and avoiding curtailment. Hydrogen can be stored, distributed and used as a fuel or feedstock in transportation, stationary or multiple industrial sectors, while creating an additional revenue streams. The presentation will also provide examples of safety, codes and standards R&D activities that FCTO is funding to enable the H2@Scale initiative. The talk will also provide a preview of the new Center for Hydrogen Safety through a collaboration between American Institute of Chemical Engineers and DOE's Pacific Northwest National laboratory, as well as multiple stakeholders worldwide.

549968 California Energy Commission Integrated Energy Policy Report

Every two years, the California Energy Commission prepares the Integrated Energy Policy Report (IEPR). The IEPR highlights the implementation of California's innovative policies and the role they have played in establishing a clean energy economy. The 2018 Update report details several key energy issues and will encompass new analyses.

551808 Effects of Elevated CO_2 Concentrations on Yields of Agricultural Crops

That elevated concentrations of CO_2 can increase plant growth has been known since at least 1804, when a scientist named De Saussure reported that peas exposed to high CO_2 concentrations grew better than control plants in ambient air (about 300 ppm at that time). Since then, thousands of experiments with elevated concentrations have been conducted, mostly in greenhouses and other types of enclosures, and almost of them similarly showed increases in growth and/or yield. Indeed, it is standard horticultural practice to enrich greenhouses to about 1000 ppm CO_2 when conditions are cool enough that so that the greenhouses do not have to be ventilated for temperature control. Then, when it was realized that the atmospheric CO_2 concentrations were increasing, scientists wondered what this portended for future plant growth and agricultural productivity in open fields. Therefore, about 30 years ago, free-air CO_2 enrichment (FACE) technology was developed that enabled the air above open-field plots to be enriched with CO_2 for entire growing seasons. Since then, FACE experiments have been conducted on cotton, wheat, ryegrass, clover, potato, grape, rice, barley, sugar beet, soybean, cassava, rape, mustard, coffee (C_3 crops), and sorghum and maize (C_4 crops). Elevated CO_2 (550 ppm from an ambient concentration of about 353 ppm in 1990) decreased evapotranspiration about 10% on average and increased canopy temperatures about 0.7°C. Biomass and yield were increased by FACE in all C_3 species, but not in C_4 species except when water was limiting. Yields of C_3 grain crops were increased on average about 19%, while other C_3 crops were increased even more.

553087 Experience with Starting and Operating a Consulting Business

In this presentation I will share my experience in starting and successfully operating a consulting business in environmental, health and safety. I started a consulting business ten years ago and since
have managed to successfully generate increasing revenue stream while maintaining good profit margins.

This talk will discuss how to carve a niche which has a demand in service business, trials and tribulations
in starting an enterprise from scratch including generating initial revenue, obtaining appropriate insurance and resources required both financial and human. I will also discuss my experience in setting
up company name and logo.

Maintaining and sustaining a small service business requires being both a" seller" and "doer". I will discuss balancing between need to sell work as well as complete technical projects. I will discuss my own
overall approach to generating steady revenue stream. I will relate my experience in maintaining cash
flow, balance sheet and other financial issues.

I will discuss some of the attitude and aptitude required to be successful in business based on my experience. There is a lot of satisfaction and a sense of accomplishment in operating a business if there
is a methodical approach with right attitude and ability to adapt to changes.

Turning a Consulting Idea into Reality

Jack Hipple, Principal

TRIZ and Engineering Training Services LLC
Tampa, FL

Abstract

Starting a consulting business is not as easy as printing business cards, posting a web site (though both are important), and waiting for the phone to ring. It involves networking, contacting friends and associates, giving pro-bono presentations at both local and national meetings, publishing, and using resources such as LinkedIn.

This presentation will review the steps needed to establish a consulting and training services consulting business from scratch and emphasizing the importance of professional association participation in turning an idea into reality.

Consulting: Is It For You?

After a 30 year industrial career, focused on both basic chemical engineering (primarily at Dow) and materials (National Center for Manufacturing Sciences, Ansell Edmont, Cabot), I was exposed to a rather unique problem solving technology known as TRIZ ("Theory of Inventive Problem Solving") and decided to look for an opportunity to work in this area. After a brief time with one of the original Russian emigrant-based consulting companies to train and learn, I decided to start my own business. Having never done this before, I encountered a number of challenges, as well as opportunities, that will be shared. They may be of some assistance and interest to those considering starting their own business.

Practical Issues

Everyone has boundaries that affect their decisions about starting an independent business and there were mine:

1. Desire to grow a business into a substantial enterprise. Many people starting consulting businesses have a goal of growing their business into multimillion-dollar enterprises, requiring hiring other employees, needing an accounting and marketing structure, and requiring decisions on employee benefits. My intention was simply to keep myself fully employed to the extent necessary to maintain a modest lifestyle.
2. Linkage with major consulting organizations. This can be a stated objective, but again, was not on my list of priorities.
3. Health insurance. The level of this concern varies greatly from state to state and is obviously affected by family size, spousal employment, and personal health issues. Given my personal health issues, this was a significant concern.

4. Family. As empty nesters with all children having completed college and employed, and our 4 daughters all married, medium term financial issues were not of concern. Obviously, it is easy to imagine other situations with college bills, teenagers at home, etc. that would have great significance.
5. Spousal support and feelings. As mentioned before, short term financial issues can be of great concern. Not all married couples see financial risk the same way and it is critical to have a frank conversation about the fact that there may be no income for several months. If a spouse is working and children are still of school age, then issues relating to child care will come up, especially if the nature of the consulting business requires travel on short notice.
6. Finances. Two key points here. First, the health issue mentioned previously. If a previous employer will or cannot allow insurance continuation (usually as a much higher cost), and a spouse is not employed by an insurance providing organization, this can be a major cost issue that needs to be considered and is part of the financial planning
7. Unpredictable events. These could be health issues but can also be unpredictable external events. Case in point was the 9/11 terrorist attack. From personal experience, and that I know of others, unless you were a security consultant, the phones stopped ringing. No one knew what was going to happen and businesses froze in place without making commitments of any sort. This can be a short-term financial crisis, and in my case, required my wife to go back to work in order to be able to maintain health insurance.
8. Travel. This again is a highly personal, and family connected decision. There are many successful consultants who do no or minimal travel. This is more common in a major metropolitan area, but that area must contain industry that has some overlap with your intended consulting focus. If heavy travel is involved, the spouse is working, and children are of school age, this can be a major, and sometimes costly, issue to deal with.

Suggestions for How to Make it Work

Since, in my case, the goal was to grow business to fully employ myself and not grow a large company, and my resources were rather limited, here were the low-cost items that allowed my business to grow into a success:

1. Business cards. Unless I know someone has died, I NEVER throw away a business card I have collected at a meeting. It costs virtually nothing (except an Email account) to send appropriate Emails to past business acquaintances telling them you have started a new business around something unique or better yet, make a phone call (a long-lost art). And after you have gotten a job or two and know you will be in a certain area over a known period, ask the right people about getting together for lunch or dinner, possibly extending the main purpose business stay for a day or two.
2. Local societies. I will address national societies next, but in almost any reasonably sized city, there are local chapters of every national engineering, business, and technical

society. And there are numerous chamber of commerce and other such groups who are surprisingly eager for speakers. It's not that the groups don't have potential speakers, but it is far more likely that a consultant has and will take the time to prepare a presentation. Paying a meeting fee of less than $50 almost always leads to follow up contacts and potential business.

3. National societies. I teach two courses for both AIChE and ASME, one in Chemical Engineering for Non-Engineers and the other in the TRIZ problem solving process. These came about, in part, by coincidence, but point out the value of maintaining a relationship, and actively participating in such societies and in their sub-groups, such as AIChE's Management Division. 15 years ago, I tried to make contact with AIChE's training group but was not terribly active in AIChE, so did not have a personal contact. I became resigned that the connection would not yield anything, but just as I was giving up the idea, I saw a course listing for TRIZ Inventive Problem Solving (my specialty) listed in the AIChE training catalog. My energy level increased to the point where I persisted, and finally talked to the training manager at the time, who was very cordial and receptive and told me that AIChE always wanted to have a back-up instructor for any course offered and gave me the opportunity to teach the next TRIZ class without ever having met me! As any good training manager would do, she came in and out of the room over the 3-day period, and at the end of the course, offered me the opportunity to be the primary instructor with no intentional switching.

   And then the next great thing happened when I was told something like, "I bet you could teach anything". After joking a bit and telling her that I was not a TV pitchman, I asked what else she had in mind. AIChE needed a backup instructor for its "Chemical Engineering for Non-Chemical Engineers" course, due its current instructor having a growing unpredictable eldercare issue that might prevent her fulfilling a commitment to teach. Though 10+ years away from having done serious ChE calculations, the idea intrigued me under the condition that I be reimbursed for going to the next public class and sitting in the back of the room and watching, after which I would decide. In the mean time I was sent a copy of the course materials to review. They did not scare me as much as I thought, so I packed my bags and got ready to fly to New Orleans on Monday morning. Late Friday afternoon, the phone rang, and I was told that the current instructor's mother had passed away and that I would be teaching the class. You can only imagine what went through my mind! I have been teaching the course ever since.

   My message here is that my major source of income, requiring little or no marketing expense on my part, has come from interactions with AIChE, thought the TRIZ class emigrated over to ASME during AIChE's financial crisis a while back. And these public classes inevitably lead to additional training opportunities, requiring little or no marketing expense on my part.

Summary

My summary of advice, given goals and situations similar to mine, is thus:

1. Make sure consulting is what you want to do
2. Focus what kind of consulting you want to do and base it on your strengths and uniqueness.
3. Make sure you and your spouse agree on this and how it will affect your lives, both personally and financially. Understand how it may affect family life. Don't leave children out of the discussion.
4. Use your past connections. Don't throw past information and resources away.
5. Prepare for 6 months of no income and consider what you will do if some unforeseen issue arises that stops your income for several months.
6. Use your local, and more importantly your national societies, to advance your business ideas and contacts.
7. NEVER throw away a business card unless you can match it with an obituary.

553059 Work in Retirement Can Yield Numerous Benefits

Several factors can induce a worker to opt to retire from the daily routine and grind of specific work assignments and requirements. Some could be health factors, family obligations, travels and fun, and the desire to impact several companies and professional workers beyond the confines of the current work sphere.

This presentation may be of interest to those who are in "retirement", to those who are contemplating "retirement", and to those who wish to continue producing and impacting companies, and others during retirement. I will be sharing with you the joys and benefits that can be derived from running a technical, engineering and optimization consultancy services.

In founding and working for Prafis Energy Solutions, I have had the time to meet family obligations. It has given me the time to re-energize my growth as I continue on my spiritual journey. "Retirement" has enabled me to work on a number of interesting projects, manage health care challenges, and welcome the recent increase in the number of my grandsons from 3 to 5. The usual travels have been done, and more recently, my book with the title of Catalytic Naphtha Reforming Process was published in 2018 by CRC Press of the Taylor Francis group.

550764 Text Mining and Analytics

Text mining seeks to extract information and insights from textual data. In particular the distillation and summary of documents is of increasing commercial interest, from news media to patents to maintenance logs.

Application areas include business analysis, scientific discovery, predictive maintenance, marketing and communications, litigation, ad cybersecurity, among others.

JBasic workflow elements will be reviewed in this talk, including: definition of the corpus (which may involve the pre-step of converting images to text), preprocessing, visualization, filtering, statistical analysis, work embeddings, and classification.

Some more advanced subjects, for example topic modeling and sentiment analysis, will be introduced and described.

Demonstrations of functionality will be presented using MATLAB.

550044 Text Analytics Applications for Material Manufacturing

Global manufacturing companies generate vast amounts of unstructured data from within their operations and enterprise resource planning (ERP) systems. These text-based data sources can be difficult to integrate into traditional reporting and analysis workflows that require querying and numerical calculations. With the advancement and ease of entry into developing text-based models and features, Dow is employing several types of text modeling methods to improve margins, reliability, safety, and our customer's experience.

Dow has developed sustainable event recording processes that provide free-form text fields to precisely record the events and actions taken, but are difficult to then aggregate into reports or understand trends. In this talk, we will share examples of how Dow has developed machine learning models to categorize text for efficient trend analysis, as well use deep learning to relate documents to one another. We will then show how leveraging these text analytics methods can deliver value through new insights and streamlined work processes.

551060 Opportunities for Text Mining in Service of Chemical Engineering

As the world has become increasingly digital, vast storehouses of data in digital form have become available both within the public sphere and within companies as proprietary resources. Over the past decades, there has been a growing awareness of the strategic value of such data to companies and to society more broadly in virtually every field - Chemical Engineering is no exception. The social web can be mined to monitor public opinion related to products and policies and how these change over time in response to news events. Documentation about work practices and events within companies can also be mined to identify ways in which company practices affect its outcomes, employee retention, and even safety concerns. Scientific literature can be mined to identify trends in research topics, common or competing findings, and emerging research interests. The news can be monitored to identify connections between environmental policies, company practices, and environmental impact.

Text mining is a field that brings together an understanding of natural language as an unstructured form of data and machine learning as a suite of modeling tools to layer structure over it. Using machine learning paradigms like probabilistic graphical models and deep learning, it is possible to architect modeling tools that are able to extract latent structure found in text. By identifying and extracting the latent structure found in text, it is possible to transform this unstructured data into structured data, at which point it is possible to apply a plethora of modeling tools to identify trends over time and even plausibly causal relations between events and states. A major goal of research in the area of text mining is to achieve robustness in the face of noisy data. With the big data revolution, new storehouses of vast amounts of textual have become available. While this data is far less well-structured than forms of language that were the target of work in natural language processing decades ago, now research on text mining applied to even the noisiest of this data (for example, from sources such as Twitter or Reddit) are commonplace.

This talk will offer a brief overview of state-of-the-art methodologies for applied machine learning, with pointers to resources and further instruction.

552694 Workshop: Managing Yourself

This session will be an **interactive workshop** that will be split into 2 parts:

1. Auditing yourself and your career: how to do a self-assessment on what skills you have, what type of role you want to have, and what skills you need to develop to get there

2. Branding yourself professionally: how to create a professional brand and market yourself

Oftentimes, a new role signifies a promise of a new future but how often do you take the time to plan out that role to prepare you for your next one? This workshop provide ideas to help you position yourself for your next role. It will also teach you how to recognize and articulate your transferable skills.

551928 Soft Skills for Young Professionals

Going from being a student to working is a big transition for most of us. It requires us to learn how to navigate the real world, and we can use all the help we can get to do it well. Soft skills are the most important secondary skill (the primary one being the vocation that you are being employed for) that can help us here, and not only in our careers but also in personal lives. This session will focus on some of the necessary soft skills regardless of the degree we have, reflect on which soft skills are transferrable and which need to be developed in order to have a rewarding time at our organizations.

551786 The *Trusted Advisor*: A Model for Career Success

The classic book *Trusted Advisor* by David H. Maister presents highly effective tactics for building and maintaining long lasting relationships as part of a successful career strategy. A Trusted Advisor is a person who has purposefully elevated their personal and professional relationships to a level where solving problems is more valuable than technical mastery. Trusted Advisors have self-confidence and driven curiosity. They believe that there is distinction between business life and personal life, but both are seamlessly interconnected.

This presentation will outline the three core skills needed to become a Trusted Advisor: earning trust, giving advice and building relationships. Effective ways to develop each of these three skills will be presented in an easy-to-understand format. Pitfalls for destroying trust will be discussed along with tactics to avoid doing so. The goal of the presentation is to outline these tactics in the form of constructive actions that you can perform daily for the rest of your career. Trusted Advisors are easy to spot: they seem confident, selfless and laser focused on helping themselves and others achieve their goals.

550124 Networking Strategies to Propel Your Career

This presentation discusses various early career strategies to build business relationships with coworkers, clients, and vendors. Develop a strong network now for any future downturn or advancing within your company.

550521 Engineering Creative Thinking

What do you think of when you hear the word Creativity? If you're like me, you often think of something artsy that you would see on Instagram or Pinterest. You also probably think creativity is a trait people are born with or without. You probably also think creativity is something you don't have to utilize in your technical jobs. However, creativity and creative thinking is a skill that can set you apart in any job. Creativity isn't about originality or thinking outside of the box as much as it is about learning how to connect ideas, growing a bigger box, and building your mental inventory of knowledge. In this session we will go over practices, tools, and tricks for how us engineers can expand our creative thinking to make us more innovative and strategic in how we approach our problem solving and our jobs.

551424 A Tale of Two Columns: Integrating Dynamic Simulation in Conceptual Design

Steady-state simulators in conjunction with standard column sizing methodologies have been used extensively in industry to design distillation columns. Steady-state simulators, when based on reliable underlying physical property data and models, are invaluable in identifying optimal designs and configurations for distillation. However, steady-state simulations do not consider the impact of hold-ups, operation in certain regimes e.g. vacuum, or commonly encountered perturbations such as feed conditions and utilities. Dynamic simulation has been used widely in industry as a tool to diagnose operability issues for operating plants and equipment and less commonly in earlier stages of process design.Dynamic simulation can also be used as an effective screening tool to identify potential control schemes. Two case studies involving new intensified designs for solvent recovery separations are presented, where dynamic simulation is used as a tool to evaluate operability in the conceptual design stage.

551020 Optimization of Mixed Smr Plant Using Advanced Modelling & Six Sigma Tools to Improve Energy Efficiency and Plant Performance: Case Study

Synthesis gas is the key feedstock for many industrial applications especially for ammonia, methanol, Oxo-alcohols etc. Most of the synthesis gas produced in the world is from natural gas using steam methane reforming (SMR). Mixed SMRs have carbon di-oxide added to the feed gas to produce synthesis gas with lower H2/CO ratio than that of conventional synthesis gas applications. As the gas reserves are depleting in the world, existing reforming plants are open to use natural gas from different sources. Different natural gas sources offer different compositions in the gas, which significantly affect the performance of the mixed steam methane reformers (SMR) in terms of reforming efficiency, energy consumption, H2/CO ratio in synthesis gas etc. Effect of feed gas disturbances in mixed SMRs is significant as these plants are operated at higher reforming temperature and lower steam to carbon ratios. In order to operate the reforming plant efficiently, optimally and smartly for all feed gas disturbances, optimization of the plant using advanced modelling & six sigma tools, precise pro-active control, continuous operations monitoring are necessary.

In this paper, an implemented optimization of mixed steam reformer in the oxo-alcohol production process case study is presented. An Aspen Plus model based study has been performed to optimize the mixed SMR plant operations for the existing feed gas composition. Then, coupled the model-based study with six sigma tools to develop the best operating conditions and zone for various scenarios. A number of contour diagrams have been developed for these scenarios using MINITAB, and response surface optimizer has been developed using response surface methodology. In this part, optimized the steam to carbon ratio, carbon dioxde to natural gas ratio, using surface response methodology using MINITAB, and the optimized conditions are translated into operator language and implemented in the systematic procedure. The implementation of the optimized conditions resulted in savings of energy 0.22 MMBTU/Ton of product and production increase of 2000 TPA.

Furthermore, a detailed dynamic model, as in the DCS of the plant, has been developed ASPEN DYNAMICS and analyzed the effects of the feed gas disturbances on the process. The analysis showed that H2/CO ratio in synthesis gas, reformer energy efficiency, methane slippage, and oxygen content in the stack gases would have serious repercussions and dynamic profiles have been developed for various feed gas compositions changes. In addition, it has been understood that a proactive advanced feed forward controls are necessary for reformer temperature, methane slippage, H2/CO ratio along with the existing feedback controls. Feed forward based control loops have been added on top of the existing controls based on the carbon number and calorific value of the feed gas. These upgradations enable the smooth operation of the plant during drastic feed gas composition changes. The added advanced process control would eliminate the situation of process upset and un-planned shutdowns, that result in improved equipment integrity and reliability.

550586 Modelling of Gas Separations By Pressure Swing Adsorption Using a Novel Steady-State Methodology

A novel pressure swing adsorption (PSA) steady-state process modeling methodology was proposed previously which greatly simplifies simulation of PSA units by using a concept of adsorption efficiency, referred to as bed-state efficiency, similar to tray efficiency in distillation. In theory, the proposed model provides a simple method for the practicing engineer to use for evaluation of PSA processes. The nature of the bed-state efficiency, however, must be elucidated before the model can see widespread acceptance. Using a combination of experimental data and dynamic simulation, an air separation unit (ASU) was investigated to gain intuition into the behavior of the bed-state efficiencies. Insights and guidance from the ASU case were then applied to the modeling of a nitrogen rejection unit (NRU) removing nitrogen from natural gas. The relationship between the bed-state efficiencies, operating conditions, and design parameters will be discussed as well as the potential for *a priori* prediction of the bed-state efficiency values.

552665 Integrated Safety Management - Taking Safety Performance to the Next Level

Gavin Towler, Vice President and Chief Technology Officer, Performance Materials and Technology, Honeywell, will provide an overview of recent advancements in risk assessment and a look ahead at the future direction for chemical and process industries. Today, risk assessment techniques play a fundamental role in process design, implementation of safety systems, maintenance, and operation management, as well as an essential tool for the development, operation and expansion of process installations. Mr. Towler will discuss the latest dynamic risk assessment approaches that support real-time decision making.

551241 Reinforcement Learning (RL) Based Data-Driven Decision Support System

Big data has transformed from a buzzword to a real value creator. In recent years there has been a rising interest in data analytics in the energy industry. The industry and academia have made significant progress in big data analytics. However, there is still a need to use the plant data to gain insights and address the issues related to abnormal situation management. An abnormal situation is defined as "a disturbance or series of disturbances in a process that causes plant operations to deviate from their normal operating state." The disturbance can be a minor near-miss to a catastrophic incident which may cause production losses, environmental damage, injuries or sometimes fatalities if not managed properly. During an abnormal situation the operator is expected to take appropriate action to bring back the process to a normal state. Diagnose and mitigation actions depend on the operator's experience, process complexity and stress levels of the operator.

This paper outlines a systematic framework for application of Reinforcement Learning (RL) during process operations and abnormal situation management. A novel method using Reinforcement Learning is established to address decision making challenges and advise operators during abnormal situations. The system is designed on the expert knowledge from operators and engineers, the knowledge available from previous operation experiences and process plant information. The application of the proposed method can be expanded by connecting the system to Operator Training Simulators (OTS) and train operators for abnormal situation management.

A SEQUENCE TO SEQUENCE TIME SERIES FORECASTING MODEL FOR BATCH PROCESSES WITH SEASONALITY

Shu Xu, Mark Nixon

Emerson Automation Solutions

Introduction

Modern chemical manufacturing processes generally fall into three categories: continuous, batch and semi-batch. Unlike its continuous counterpart where the raw materials are constantly fed into the system and processed into an intermediate or final product continuously, a batch process receives and processes the feedstock by a series of discrete steps. Such a manufacturing process has been widely used in semiconductor, pharmaceutical, specialty chemical and other industries. For example, the fermentation process used in brewing industry. Because the whole process consisting of many operating steps can hardly be operated in a steady state, it poses tougher challenges for both equipment and operators to assure product quality in many batch runs. One specific challenge is related to seasonality which stems from equipment aging[1], periodic maintenance[2], catalyst decay [3] and so on. Typical examples of such a behavior involve activity coefficients of reactor catalysts and fouling factors in heat exchangers[4]. Process engineers are challenged to forecast such a seasonality lurking in the system so that they can become proactive in identifying and preventing potential problems.

In this paper, we built a sequence-to-sequence model to forecast the seasonality simulated in the benchmark Penicillin bio-reactor process. The results obtained by such a sequence-to-sequence model are compared with those generated from a linear partial least square (PLS) method and a nonlinear support vector machine (SVM) regression method.

The remainder of this paper is organized as follows. Section 2 briefly introduces methodology information regarding time series forecasting, long short-memory (LSTMs) networks, and an associated sequence-to-sequence model. Section 3 provides case study details including the benchmark Penicillin bio-reactor simulator and simulation conditions. Section 4 summaries results obtained in the case study. We conclude the paper and discuss future works in Section 5.

Methods

Time series forecasting

During batch operations, abundant informative data can be collected in the form of time series with regular sampling periods [5]. As a result, forecasting methods in time series analysis, which take advantage of dependence between observations in the time dimension to predict the future, can be applied in process seasonality forecast [6], for example, classical methods include autoregression, autoregressive integrated moving average (ARIMA), simple exponential smoothing (SES), etc. In addition, time series forecasting can also be reframed as a supervised learning problem. By reframing the time series data via a lag method, various linear and nonlinear supervised learning algorithms can be used, including linear regression, support vector machine, neural networks, or a sequence-to-sequence model with long short-term memory (LSTM) networks as discussed in this paper.

LSTM network

Long short-term memory (LSTM) networks belong to the category of recurrent neural networks (RNNs, see Figure 1) where loops and chains exist to allow information to persist. In Figure 1, an input x_t is fed into a regular neural network block A which produces an output h_t. An RNN distinguishes itself from its regular counterpart in that each neural network unit in the RNN can pass information either to itself in a loop or to a successor to form a chain. Although a normal RNN can track the short-term dependence between observations in the time series, it struggles to learn the "long-term" connections [7]. To overcome such a challenge, the LSTM network is proposed[8] (see Figure 2).

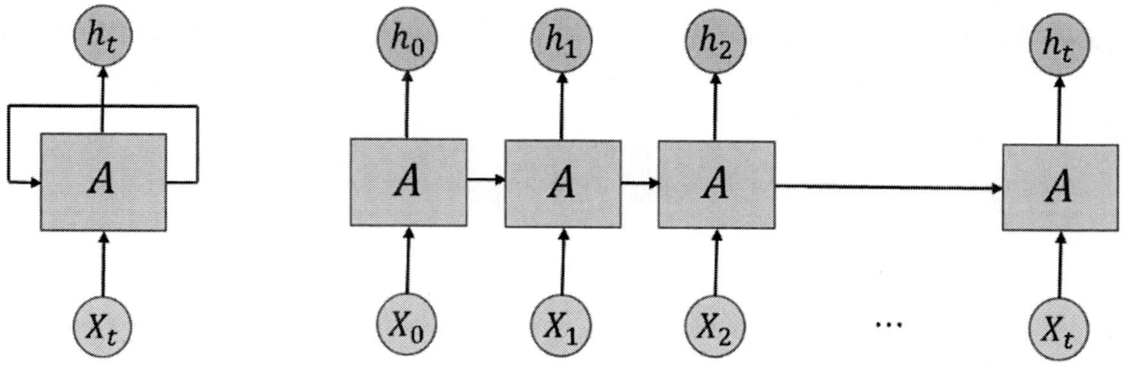

(a) An RNN with loop (b) An unrolled RNN

Figure 1 Recurrent neural network demonstration

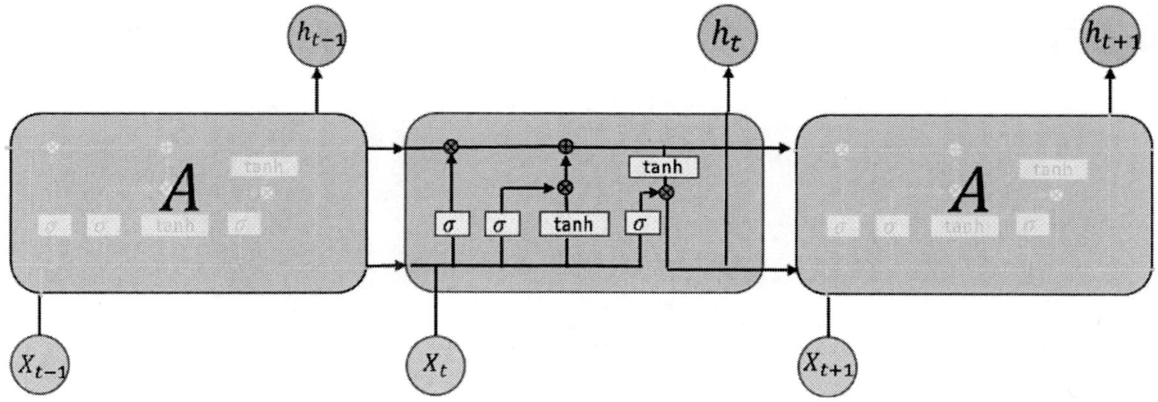

Figure 2 LSTM network demonstration

Although the general form of the LSTM network is akin to that of a regular RNN, its repeating cell each contains four neural network layers interacted in a way designed specifically for remembering information for long periods of time. A detailed description of the structure of each cell is shown as follows.

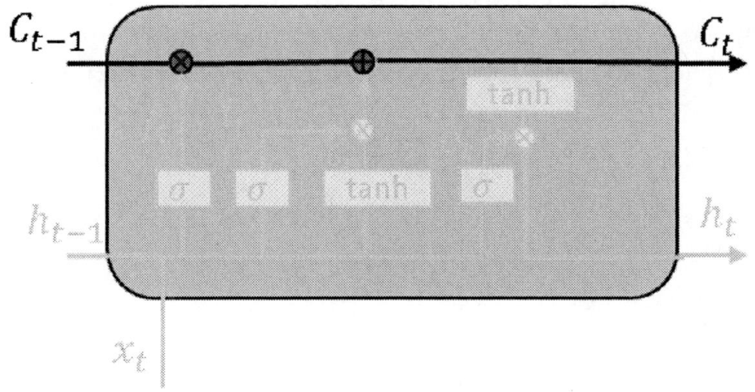

Figure 3 Cell state in an LSTM cell

As illustrated in Figure 3, a horizontal line running across the top of cell along with only linear pointwise operation indicates potentially a free flow of information without any cell status changes. In addition, the amount of cell status information gain or loss is controlled by the following gates in Figure 4, which are comprised of a sigmoid neural net layer and a pointwise multiplication operation. Because the sigmoid layer outputs a number between 0 and 1, it regulates how much information should be let through. While a value of 0 indicates closing the gate, a value of 1 means letting the information flow through freely.

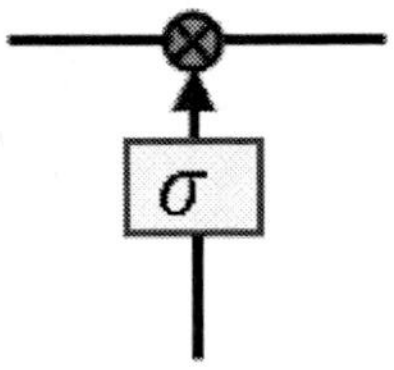

Figure 4 Gate in an LSTM cell

Figures 5~8 provide a step-by-step walk-through of the four-layer structure inside each LSTM cell. As shown in Figure 5, the first layer named the "forget gate layer" decides whether to keep the previous cell state C_{t-1} or not. For example, if the forgetting factor f_t calculated by gate Equation (1) equals 0, the previous cell state is completed discarded (or "forgotten") after an element-wise multiplication.

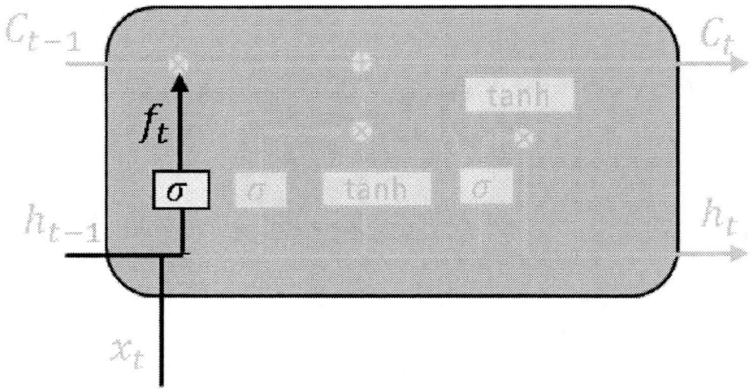

Figure 5 Forget gate layer in an LSTM cell

$$f_t = \sigma\left(W_f \cdot [h_{t-1}, x_t] + b_f\right) \qquad (1)$$

The next state update layer consists of two sub-layers, out of which the "input gate layer" decides whether we need to update the cell state with the new value produced by the "new value layer", as shown in Figure 6. The associated gate and value equations are listed in Equations (2) and (3) respectively.

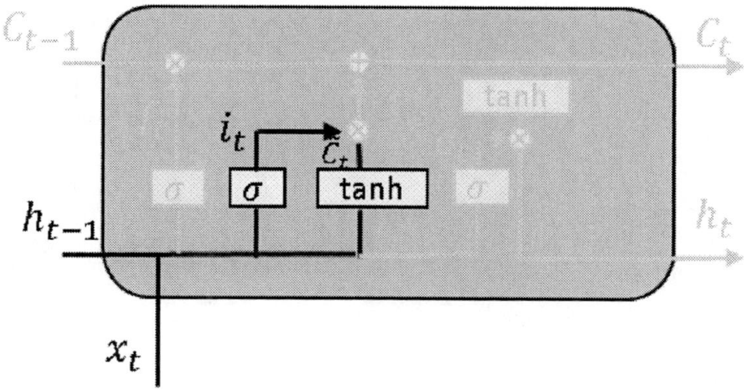

Figure 6 State update layer in an LSTM cell

$$i_t = \sigma(W_i \cdot [h_{t-1}, x_t] + b_i) \tag{2}$$

$$\tilde{C}_t = tanh(W_C \cdot [h_{t-1}, x_t] + b_C) \tag{3}$$

Now since we have both the previous cell state C_{t-1} and the new state update value $\tilde{C}_t$, the new cell state C_t can be calculated by Equation (4), and the diagram of state update is exhibited in Figure 7.

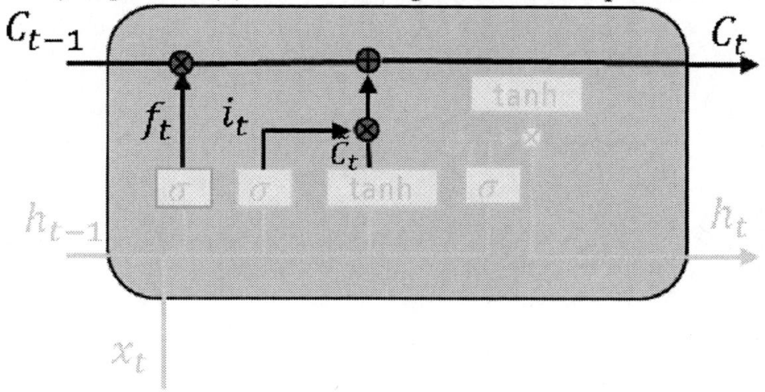

Figure 7 State update operation in an LSTM cell

As shown in Equation (4), the new state C_t equals the sum of an old state C_{t-1} multiplied by the forgetting factor f_t, and a new state update $\tilde{C}_t$ scaled by a factor i_t.

$$C_t = f_t * C_{t-1} + i_t * \tilde{C}_t \tag{4}$$

Finally, the output of the cell is decided by the output layer shown in Figure 8, including a sigmoid gate which decides whether we need to deliver a filtered version of the cell state C_t, the h_t, and associated equations are shown in Equations (5) and (6).

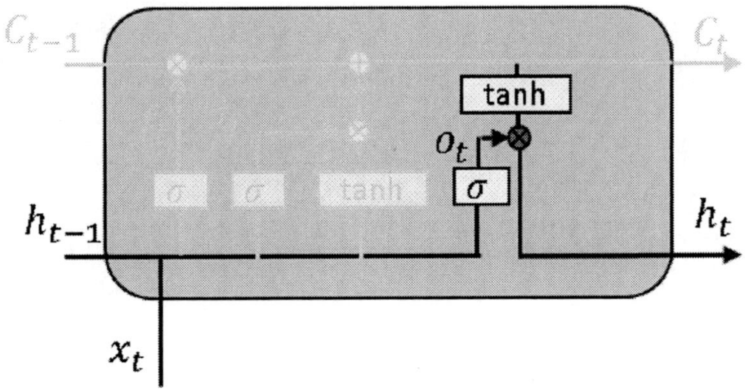

Figure 8 Output layer in an LSTM cell

$$o_t = \sigma(W_o \cdot [h_{t-1}, x_t] + b_o) \quad (5)$$

$$h_t = o_t * tanh(C_t) \quad (6)$$

In our study, a sequence to sequence (seq2seq) model with LSTM cells is used for time series forecasting. The model includes an encoder converting an input sequence X_t to corresponding state vectors which contain transformed context for each input point x_t, and a decoder which takes as input such context vectors, internal hidden states to generate an output sequence Y_t.

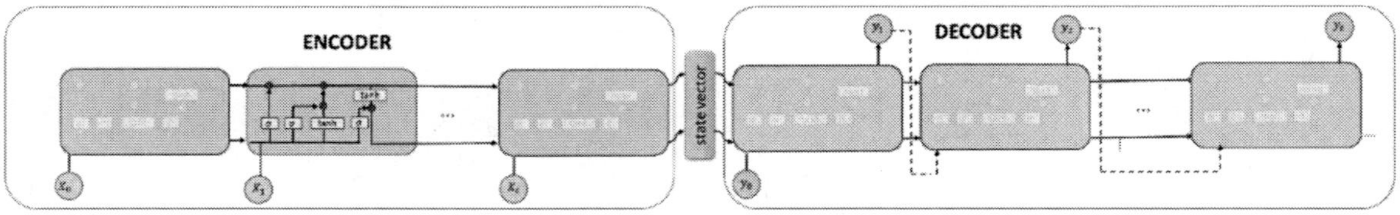

Figure 9 A seq2seq model with LSTM cells

Case study

The Benchmark Penicillin Bio-reactor simulator

A lumped-parameter dynamic batch-fed Penicillin model developed by Birol et al. [9] is used as the basis for current study. The original MATLAB code for the simulation program can be downloaded from Dr. Cinar's website (www.chee.iit.edu/~cinar/software.htm), and modifications have been made on the key kinetic parameter to simulate the batch seasonality. The main state variables included in this simulation are: reactor temperature, biomass volume, pH, biomass concentration, substrate concentration, penicillin concentration, dissolved oxygen concentration, carbon dioxide concentration. The full model is provided in the Appendix. The ODE23t stiff solver provided in MATLAB is used to simulate the process.

The general process flow diagram of the Penicillin reactor is shown in Figure 10. In the initial growth phase, the fermenter is initiated with a small volume of bacteria culture and substrate (glucose and oxygen). The reactor temperature is kept at 298 K (room temperature) to initiate the growth of the Penicillin producing bacteria. After approximately 50 hours, the initial substrate is depleted; the reactor then enters the penicillin production phase. In this phase, the fermenter is operated in semi-batch mode. Continuous flows of substrate (through the substrate tank) and oxygen (through the aerator in the bottom of the fermenter) promote the production of penicillin, carbon dioxide and other by-products.

Furthermore, two PID controllers regulate the acidity level (pH ~5) and temperature (298 K) of the fermenter to promote Penicillin production.

Figure 11 plots the resulting trajectory of a single batch. The sampling frequency is set to once every 30 minutes, mimicking that of real plant data. In addition, additional output variables of the simulation include agitator power, aeration rate, substrate flow rate, cooling water flow rate, heat generation. These variables are either manipulated inputs, or measured variables of the PID control loop.

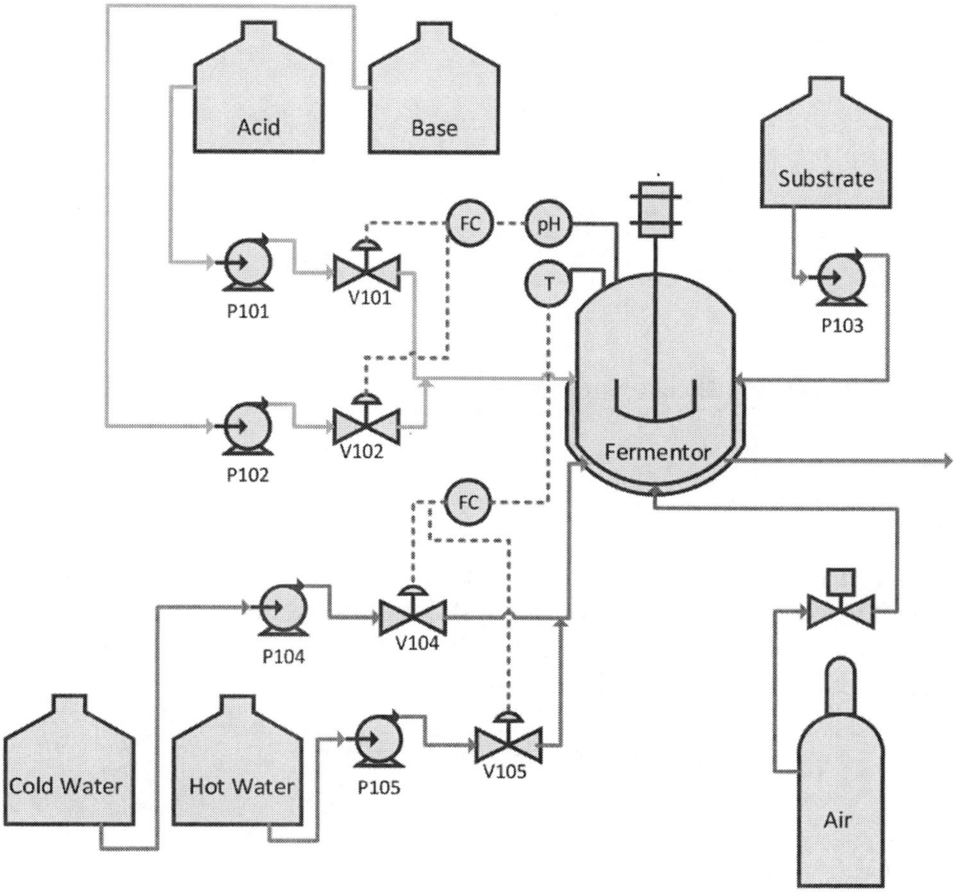

Figure 10 Process flow diagram of the batch-fed penicillin process

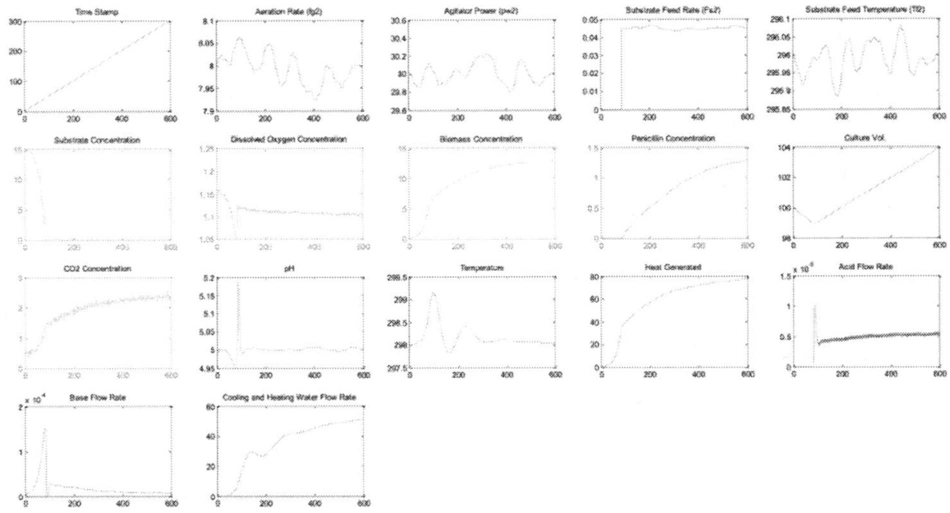

Figure 11 Example batch trajectories from a simulation run of the penicillin reactor

Simulation Conditions

To introduce process seasonality, the specific growth rate μ in the biomass growth equation (see Equation (7)) is configured to follow a first-order exponential decay with time constant set to 150 batches. The decay of the specific growth rate μ starts from the nominal value of 0.092 till the terminal value of 0.048. As shown in Figure 12, scheduled maintenance (occurs once every 100 batch cycles) of the reactor will reset the specific growth rate back to its nominal value.

$$\frac{dX}{dt} = \mu X - \frac{X}{V}\frac{dV}{dt} \tag{7}$$

Table 1 lists the simulation conditions for our current study. A total of 600 batches with seasonality corresponding to six cycles in Figure 12 are simulated, out of which 400 batches are used for training, and the remaining 200 are used for testing. All the data goes through batch-wise unfolding as shown in Figure 13, where 3D batch data are flattened batch-wise into a 2D matrix. After unfolding, data from the ith batch are squeezed into the ith row of the matrix, and such a row is comprised of values of variable j measured at time k. After scaling and mean centering of the flattened matrix, a lag of 10 batches are used in matrix reformatting for supervised model building. In other words, we use prior 10 batches of data to forecast the next batch.

All the time series forecasting models, including the sequence-to-sequence (seq2seq), support vector machine (SVM) regression, and partial least squares (PLS), are built in python, where libraries such as tensor flow, sklearn, pandas, and numpy are used.

Table 1 Simulation conditions for batch data generation

Case	Training	Testing
	400 batches with seasonality	200 batches with seasonality

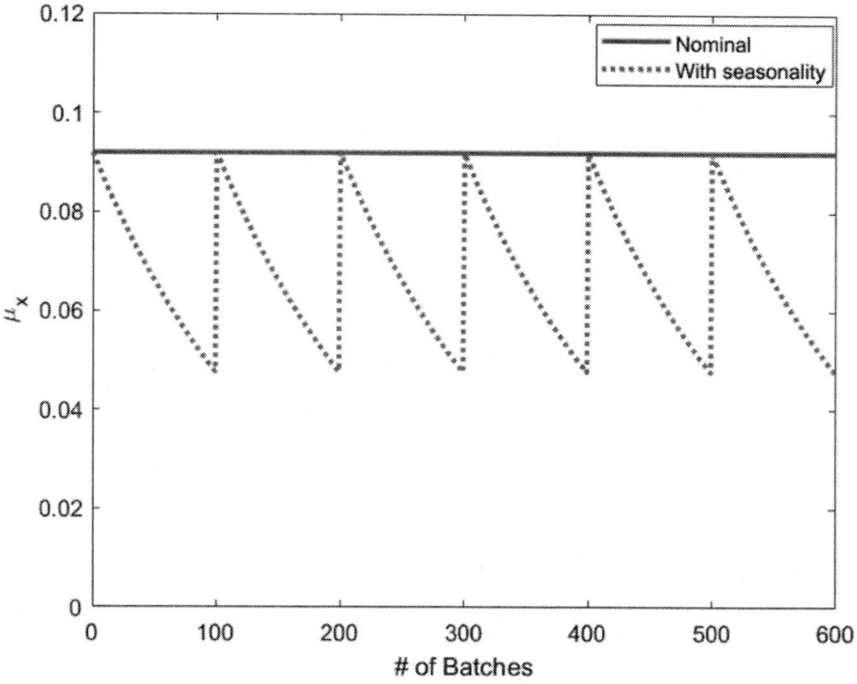

Figure 12 Seasonality in specific growth rate of bacteria culture over six cycles (decay and reset)

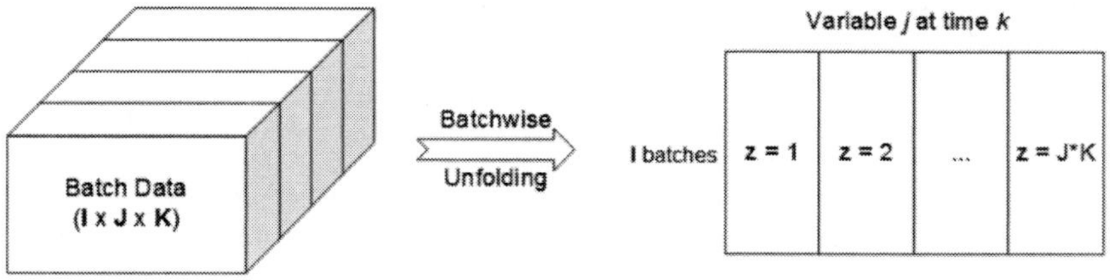

Figure 13 Batch-wise unfolding scheme in time series forecasting

Results and Discussions

In this study, we use the root mean square error of forecasting calculated in Equation (8) to evaluate the forecasting performance, and the associated results are listed in Table 2. It is observed that the seq2seq model outperforms the other two supervised learning methods in terms of forecast accuracy of batch seasonality.

$$RMSEF = \sqrt{\frac{\sum_{i=1}^{n}(y_i - \hat{y}_i)^2}{n}} \tag{8}$$

Table 2 RMSEF values of forecasting models

Forecasting models	RMSEF
Seq2seq	3.41

SVM	4.40
PLS	4.46

The actual forecasting results are illustrated in Figures 14~16. It is observed that the nonlinear models including the seq2seq and SVM obtain smoother forecasting results than the linear PLS model. In addition, the seq2seq model excels in forecasting the exact batch where the production of penicillin rises abruptly due to reset of bacteria growth rate.

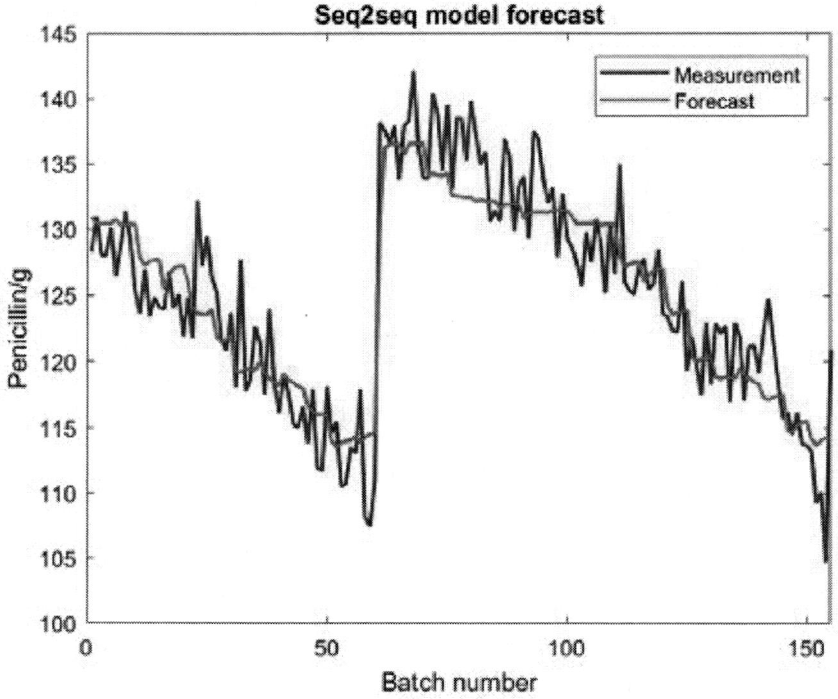

Figure 14 Seq2seq model forecasting results

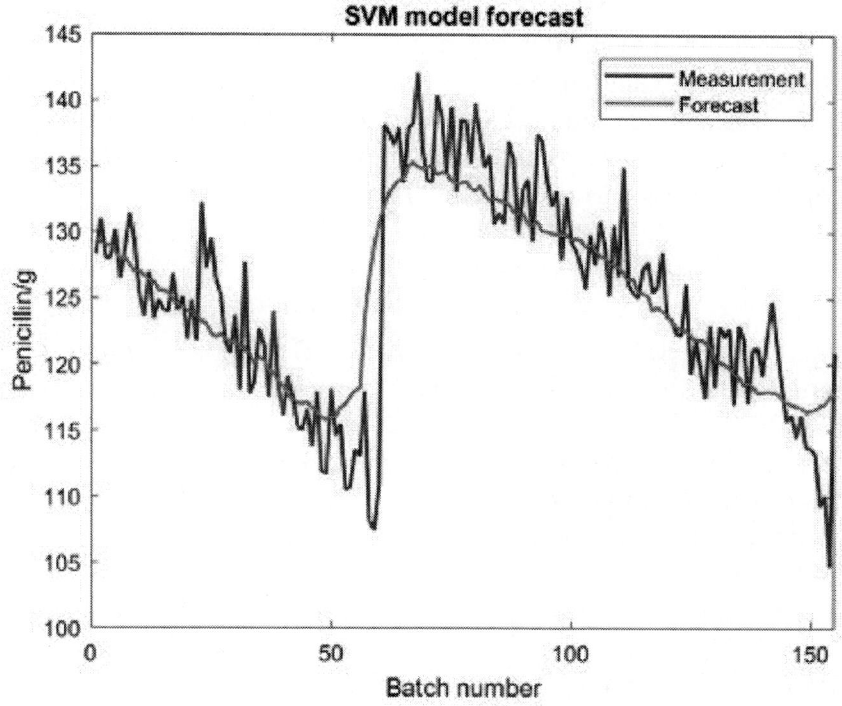

Figure 15 SVM model forecasting results

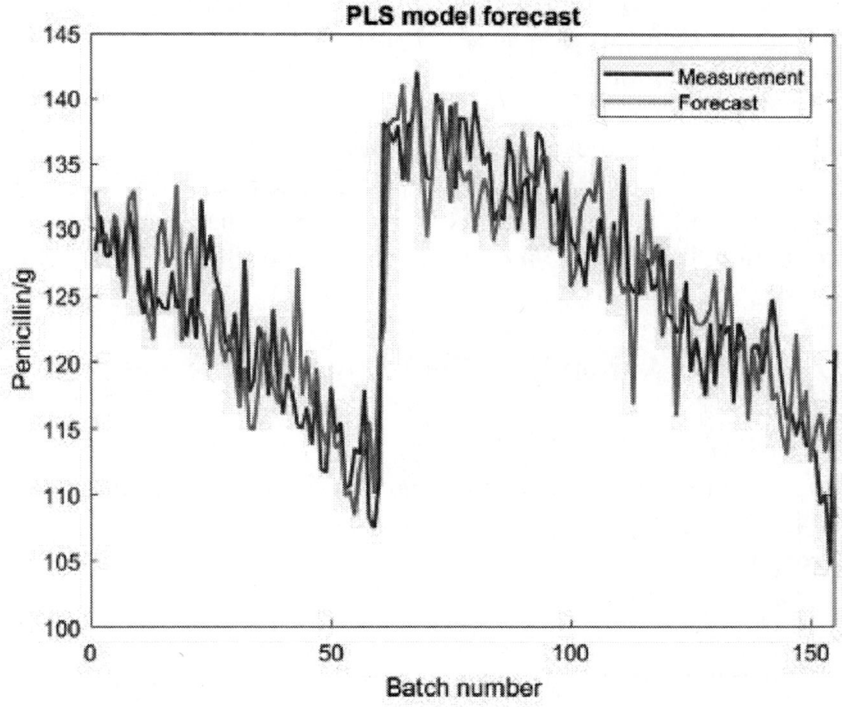

Figure 16 PLS model forecasting results

Conclusions and future work

In current work, we demonstrate the adoption of the sequence to sequence learning framework to overcome the challenge of batch seasonality forecast. A time series forecasting model is built in Tensor

Flow, and it is tested on the data generated from the benchmark Penicillin bio-reactor simulator. The model successfully forecasts the cyclic change of penicillin production rate with the periodic decay of the biomass growth rate. Such results illustrate the successful application of data analytics to wrestle with real problems in chemical manufacturing within the Digital Transformation era.

In the future, we will test such a seq2seq framework on more industrial data and compare its performance with other time series forecasting methods.

References

[1] C. Clarotti, A. Lannoy, S. Odin, and H. Procaccia, "Detection of equipment aging and determination of the efficiency of a corrective measure," *Reliab. Eng. Syst. Saf.*, vol. 84, no. 1, pp. 57–64, Apr. 2004.

[2] M. Ji, Y. He, and T. C. E. Cheng, "Single-machine scheduling with periodic maintenance to minimize makespan," *Comput. Oper. Res.*, vol. 34, no. 6, pp. 1764–1770, Jun. 2007.

[3] B. W. Wojciechowski, "A theoretical treatment of catalyst decay," *Can. J. Chem. Eng.*, vol. 46, no. 1, pp. 48–52, 1968.

[4] H. Müller-Steinhagen, *Heat exchanger fouling: mitigation and cleaning techniques*. IChemE, 2000.

[5] S. J. Qin, "Process Data Analytics in the Era of Big Data," *AIChE J.*, vol. 60, no. 9, pp. 3092–3100, 2014.

[6] G. E. P. Box, G. M. Jenkins, G. C. Reinsel, and G. M. Ljung, *Time series analysis: forecasting and control*. John Wiley & Sons, 2015.

[7] Y. Bengio, P. Simard, P. Frasconi, and others, "Learning long-term dependencies with gradient descent is difficult," *IEEE Trans. neural networks*, vol. 5, no. 2, pp. 157–166, 1994.

[8] S. Hochreiter and J. Schmidhuber, "Long short-term memory," *Neural Comput.*, vol. 9, no. 8, pp. 1735–1780, 1997.

[9] G. Birol, C. Ündey, S. J. Parulekar, and A. Çınar, "A morphologically structured model for penicillin production," *Biotechnol. Bioeng.*, vol. 77, no. 5, pp. 538–552, 2002.

[10] G. Birol, C. Ündey, and A. Cinar, "A modular simulation package for fed-batch fermentation: penicillin production," *Comput. Chem. Eng.*, vol. 26, no. 11, pp. 1553–1565, Nov. 2002.

Appendix – Penicillin Batch Reactor Model

Modeling Equations

The modeling equations and associated parameters, constants are taken from Birol et al. [10] and are simply provided here for completeness and convenience.

Biomass Growth

$$\frac{dX}{dt} = mX - \frac{X}{V}\frac{dV}{dt}$$

where μ is the specific growth rate and contains the effects of environmental variables (pH and Temperature) as well as carbon source (glucose) and oxygen in its kinetic expression.

Penicillin Production:

$$\frac{dP}{dt} = m_{pp}X - KP - \frac{P}{V}\frac{dV}{dt}$$

where μ_{pp} is the specific penicillin production rate containing biomass, carbon source (glucose) and oxygen concentrations in its kinetic expression. Here, the hydrolysis of penicillin is also considered and expressed as a first order rate expression with a rate constant K.

Substrate Utilization:

Glucose:

$$\frac{dS}{dt} = - \frac{m}{Y_{xs}}X - \frac{m_{pp}}{Y_{ps}}X - m_X X + F - \frac{S}{V}\frac{dV}{dt}$$

Oxygen:

$$\frac{dC_L}{dt} = - \frac{m}{Y_{xo}}X - \frac{m_{pp}}{Y_{po}}X - m_0 X + K_{la}\left(C_L^* - C_L\right) - \frac{C_L}{V}\frac{dV}{dt}$$

Here, K_{la} is taken to be a function of agitator power input and flow rate of oxygen as suggested by Bailey and Ollis (1986).

Carbon Dioxide Production:

$$\frac{dCO_2}{dt} = a_1 \frac{dX}{dt} + a_2 X + a_3$$

The rate expression for CO_2 has been taken from the earlier work of Montague et al. (1986).

Effect of pH

$$\frac{d\left[H^+\right]}{dt} = \gamma\left(\mu X - \frac{FX}{V}\right) + \left[\frac{-B + \sqrt{\left(B^2 + 4\times10^{-14}\right)}}{2} - \left[H^+\right]\right]\frac{1}{\Delta t}$$

Where $B = \frac{\left[10^{-14}/\left[H^+\right] - \left[H^+\right]\right]V - C_{a/b}\left(F_a + F_b\right)\Delta t}{V + \left(F_a + F_b\right)\Delta t}$, F_a and F_b represents acid and base flow rates in l/h, respectively.

Volume Change

$$\frac{dV}{dt} = F + F_{a/b} - F_{loss}$$

where $F_{a/b}$ is the flow of acid/base, F_{loss} accounts for the evaporative losses during fermentation:

$$F_{loss} = V\lambda\left(e^{5(T-T_o)/(T_v-T_o)} - 1\right)$$

Heat of Reaction

$$\frac{dQ_{rxn}}{dt} = r_{q1}\frac{dX}{dt}V + r_{q2}XV$$

where *r* are yield coefficient associated with the exothermic microbial reactions.

Energy Balance

$$\frac{dT}{dt} = \frac{F}{s_f}\left(T_f - T\right) + \frac{1}{V\rho c_p}\times\left[Q_{rxn} - \frac{aF_c^{b+1}}{F_c + \left(aF_c^b / 2\rho_c c_{pc}\right)}\right]$$

PID controller

A discretized PID control law has been used in both the temperature and pH controls. E_N denotes the error at time step N and E_{N-1} denotes the error at time step N-1.

$$\Delta MV_N = K_c\left(E_N - E_{N-1} + \frac{\Delta t}{\tau_1}E_N - \frac{\tau_d}{\Delta t}\times\left(CV_N - 2CV_{N-1} + CV_{N-2}\right)\right)MV_N$$

Overall, there are total of 15 differential equations which are solved simultaneously. All the parameters are taken from literature or assigned values which are physically meaningful in the absence of available literature. Keeping in mind that most of the parameters are strain specific and also are affected by the nature of the substrate and the environmental conditions like pH and temperature, the user should be aware of the fact that the parameters used in this simulator are average values rather than specific to a certain strain or an environmental condition.

Initial Conditions, Kinetic and Controller Parameters

Initial conditions	
Substrate concentration: S (g/l)	15
Dissolved oxygen concentration: C_L ($= C_L^*$ at saturation) (g/l)	1.16
Biomass concentration: X (g/l)	0.1
Penicillin concentration: P (g/l)	0
Culture volume: V (l)	100
Carbon dioxide concentration: CO_2 (mmol/l)	0.5
Hydrogen ion concentration: $[H^+]$ (mol/l)	$10^{-5.1}$
Temperature: T (K)	297
Heat generation: Q_{rxn} (cal)	0
Kinetic parameters and variables	
Feed substrate concentration: s_f (g/l)	600
Feed flow rate of substrate: F (l/h)	
Feed temperature of substrate: T_f (K)	298
Yield constant: $Y_{x/s}$ (g biomass/g glucose)	0.45
Yield constant: $Y_{x/o}$ (g biomass/g oxygen)	0.04
Yield constant: $Y_{p/s}$ (g penicillin/g glucose)	0.90
Yield constant: $Y_{p/o}$ (g penicillin/g oxygen)	0.20
Constant: K_1 (mol/l)	10^{-10}
Constant: K_2 (mol/l)	7×10^{-5}
Maintenance coefficient on substrate: m_x (per h)	0.014
Maintenance coefficient on oxygen: m_o (per h)	0.467
Constant relating CO_2 to growth: α_1 (mmol CO_2/g biomass)	0.143
Constant relating CO_2 to maintenance energy: α_2 (mmol CO_2/g biomass h)	4×10^{-7}
Constant relating CO_2 to penicillin production α_3 (mmol CO_2/l h)	10^{-4}
Maximum specific growth rate: μ_x (per h)	0.092
Contois saturation constant: K_x (g/l)	0.15
Oxygen limitation constant: K_{ox}, K_{op} (no limitation)	0
Oxygen limitation constant: K_{ox}, K_{op} (with limitation)	2×10^{-2}, 5×10^{-4}
Specific rate of penicillin production: μ_p (per h)	0.005
Inhibition constant: K_p (g/l)	0.0002
Inhibition constant for product formation: K_I (g/l)	0.10
Constant: p	3
Penicillin hydrolysis rate constant: K (per h)	0.04
Arrhenius constant for growth: k_g	7×10^3
Activation energy for growth: E_g (cal/mol)	5100
Arrhenius constant for cell death: k_d	10^{33}
Activation energy for cell death: E_d (cal/mol)	50000
Density × heat capacity of medium: ρ C_p (per l °C)	1/1500
Density × heat capacity of cooling liquid: ρ_c C_{pc} (per l °C)	1/2000
Yield of heat generation: r_{q_1} (cal/g biomass)	60
Constant in heat generation: r_{q_2} (cal/g biomass h)	1.6783×10^{-4}
Heat transfer coefficient of cooling/heating liquid: a (cal/h °C)	1000
Cooling water flow rate: F_c (l/h)	
Constant: b	0.60
Constants in K_{la}: α, β	70, 0.4
Constant in F_{loss}: λ (per h)	2.5×10^{-4}
Proportionality constant: γ (mol $[H^+]$/g biomass)	10^{-5}
Controller parameters (PID)	
pH: (base) K_c, τ_I: (h), τ_d: (h)	8×10^{-4}, 4.2, 0.2625
pH: (acid) K_c, τ_I: (h), τ_d: (h)	1×10^{-4}, 8.4, 0.125
Temperature: (cooling) K_c, τ_I: (h), τ_d: (h)	70, 0.5, 1.6
Temperature: (heating) K_c, τ_I: (h), τ_d: (h)	5, 0.8, 0.05

Progress Toward Machine Learning Method for Corrosion Surveillance of Nuclear Materials Storage Containers

Bruce Hardy[1], Jason Bakos[2], Taylor Clingenpeel[2], Anna dEntremont[1], Michael Martinez-Rodriguez[1], Brenda Garcia-Diaz[1] and Lindsay Roy[1],

1. Savannah River National Laboratory, Aiken, SC,
2. Computer Science and Engineering, University of South Carolina, Columbia, SC

Introduction

Within the DOE complex, 3013 canisters are used to store Pu-bearing waste. Wall penetration due to corrosion, specifically stress corrosion cracking, is considered to be the most likely cause of failure over the 50-year lifetime of the canisters.

The 3013 canister system consists of three nested canisters: an innermost "convenience can", containing the Pu-bearing material, that is sealed inside an inner 3013 canister serving as the primary containment, which is, in turn, placed in an outer 3013 canister that serves as a secondary containment, see Figure 1. The safety basis for the 3013 system, mandates that there is no gas leakage from the inner 3013 can. The top of the inner 3013 canister is sealed by welding, and the weld-affected region is called the Inner Can Closure Weld Region (ICCWR), see Figure 2. Over time, halides contained in the Pu-bearing material can be released from the convenience can into the gas space within the inner canister. In simulated exposures using test coupons, exposure to the halides have been found to produce corrosion preferentially in the ICCWR.

Figure 1 3013 nested canister system.

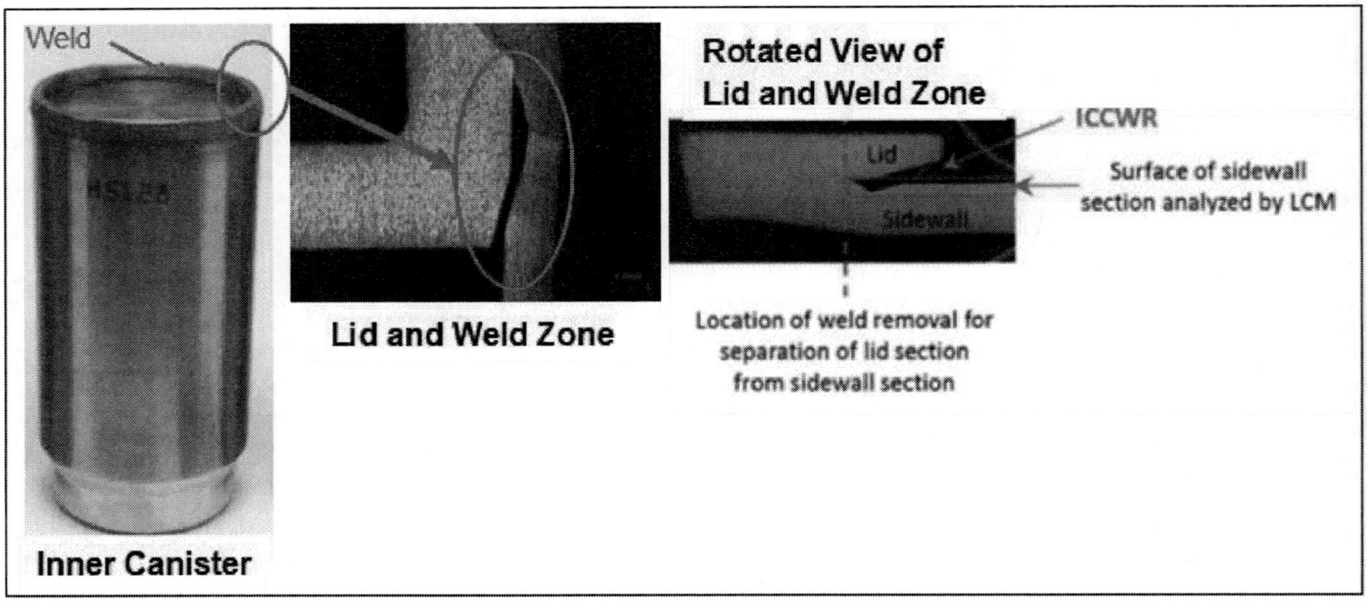

Figure 2 Inner Can Closure Weld Region (ICCWR) and sectioned canister.

Corrosion inspection is part of the Material Identification and Surveillance (MIS) program for long term storage of plutonium-bearing materials in the DOE complex. It is an objective of the MIS program to develop evaluation methodologies and metrics for assessment of corrosion. Currently, the principal method for corrosion inspection is image data obtained using a Laser Confocal Microscope (LCM). In the inspection process, randomly selected canisters are cut into sections and examined via LCM imaging. The LCM produces binary data files that contain surface height data, RGB color data (both optical and [optical + laser intensity]) and laser intensity data, yielding 8 layers of data per pixel. There are approximately 6000 images produced per can with each image measuring 1024 by 768 pixels, yielding 786,432 pixels per image, with the aforementioned 8 layers of data for each pixel. The scans are used to analyze statistics for features associated with corrosion. After establishing the statistical protocol, SRNL must work through a 5-year backlog of images while continuing to receive new can images for analysis. Initially, manual inspection was attempted, but given the vast quantity of data and its complexity, an effort was undertaken to implement numerical methods including computer vision and machine learning algorithms.

Processing of LCM Data

Corrosion is strongly, but not exclusively, associated with surface pitting and cracking. Coloration, shapes and patterns of surface features are also indications of corrosion. Conversely, not all pits and surface lesions are the result of corrosion, some are artifacts of fabrication, impact, scoring or other non-corrosion events. In manual observations, corrosion is identified via the combined properties of pit depth, area, edge contour, color and clustering. Examples of progressively degraded surfaces are shown in Figure 3. Surface data, representing physical characteristics, are extracted from the binary LCM files and used to identify features of interest. Surface characteristics, indicative of corrosion, are obtained by applying mathematical operations to the features. The mathematical operations are those invoked for computer vision and include, but are not limited to, gradients, correlations and filters.

In the most simplistic application, image data is used directly by flagging parameters, such as pit depth and cracking, that are associated with the incipience or presence of corrosion. Flagged surface features are then examined manually for final determination of whether corrosion is has occurred. While this approach greatly improves the efficiency of the examination process, it is still be very time consuming. A more sophisticated approach is to assess the data using machine learning algorithms to identify corrosion without manual intervention. The latter is the goal of the work discussed in this paper.

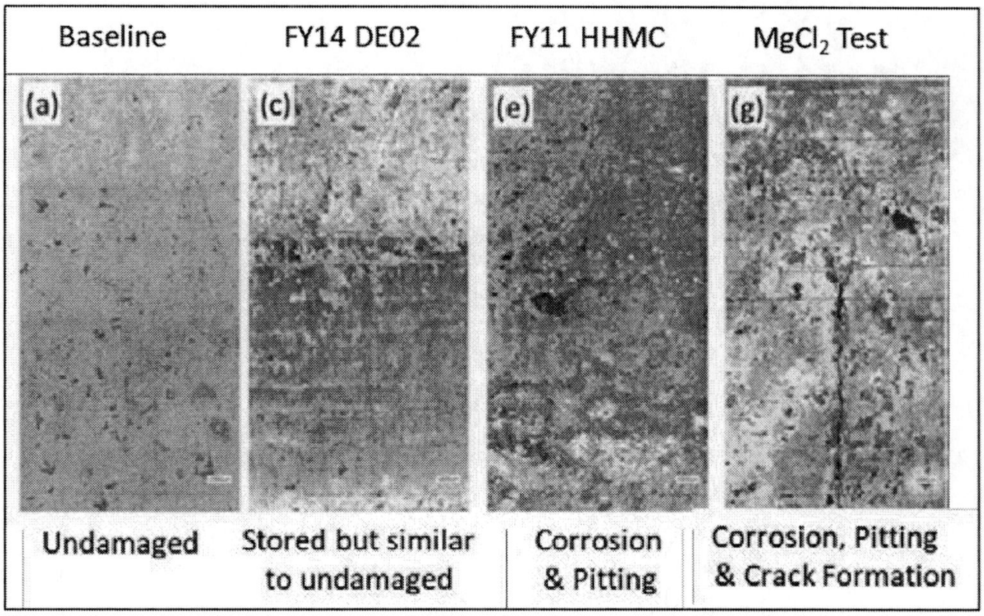

Figure 3 Optical images ranging from undamaged to severely damaged ICCWR. The HHMC is a canister exposed to high humidity and the $MgCl_2$ test was a sample exposed to boiling $MgCl_2$ to accelerate corrosion.

It goes without saying that any approach to numerical evaluation requires the binary LCM data to be converted back to physical properties; namely height, RGB color and grayscale image values. Recovery of the physical properties can be accomplished using software such as Matlab®[1] or Gwyddion[2]. However, the LCM data requires additional processing to make it useful for identification of corrosion. Explaining this requires a description of the way that a sample is examined by the LCM. For example, consider the computation of the local surface height, which is a measure of the depth of pits, or the height of protrusions, relative to the neighboring surface. In the LCM measurement process, the inner canister is first sectioned and placed into a sample holder, see Figure 4. The samples have curvature, due to the cylindrical shape of the canister, and tilt, due to the way they are mounted in the holder. LCM height measurements provide the absolute surface height, which is the combined result of tilt, curvature, pits cracks, etc. Clearly, for corrosion associated features, such as pits, the height data of interest is taken relative to the neighboring surface. However, the height of the surface neighboring the feature is the average of the local surface height, which is a function of tilt and curvature effects. A GUI was developed using Matlab® App Designer to simplify the input and processing of the LCM data. Through the GUI, the user is able to: generate plots, images and histograms; flag surface structures exceeding input thresholds; and select either single images or, select the number of images to be stitched into a mosaic, for interrogation.

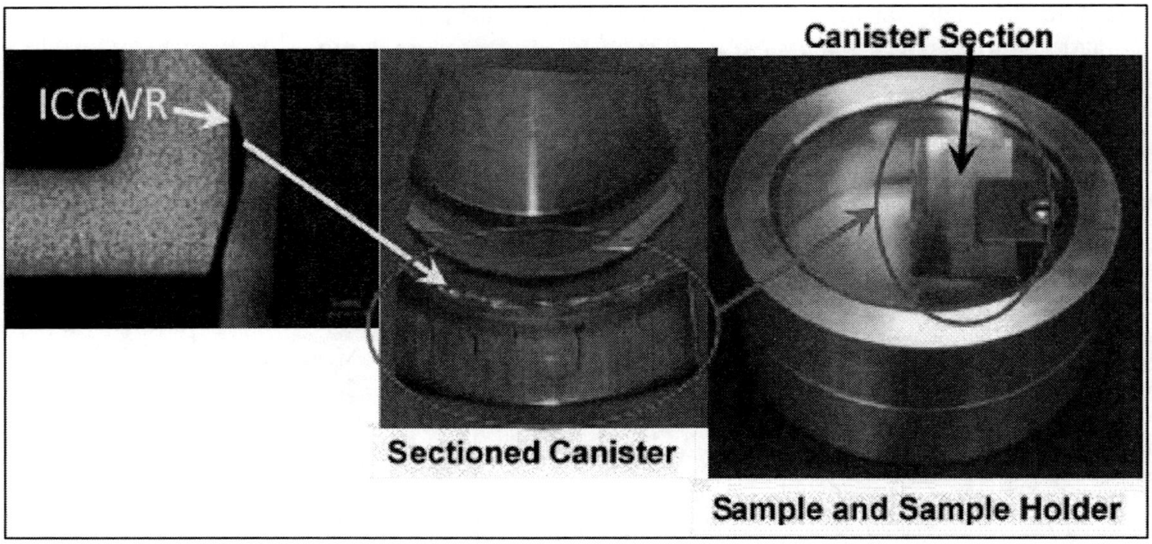

Figure 4 Sectioned inner canister and loading of section in sample holder

To obtain the general physical surface features, LCM data is extracted from large binary files, with software written to convert the data to physical attributes (i.e. height, color and grayscale values; all as functions of location on a projection of the canister surface onto a plane). Since the sample surface is not flat, the software computes the mean surface height by applying the user's choice of either a polynomial regression of varying degree or a spectral method (SMOOTHN[3]) The mean surface height is then used as a reference to determine the depth of depressions and protrusions. The user interface for the software permits selective downloading of the binary data and interrogation of the attributes. User input thresholds are used to flag attributes of interest. Physical attributes are used to compute features associated with corrosion, such as the volumes, shape and cross-sectional areas of depressions, pits, cracks and prominences. The features are accessible to permit further interrogation of the surface data through the user interface, as well as for use in machine learning algorithms.

The LCM output data consists of a collection of images, each of which is a very small part of the overall sample surface area. Individual images are rectangular, composed of 1024 by 768 pixels, with each pixel being a square that is 694.92nm long on each side. To obtain a large-scale perspective of the surface, the images must be stitched together to form a mosaic. A mosaic composed of stitched images is important when dealing with features that span many images, permitting a more complete overview of surface structure. However, a difficulty arises due to the way the LCM outputs height data on an image-wise basis. For each image, the LCM measures surface height relative to an arbitrary reference. This means that the absolute heights reported by the LCM can differ discontinuously between adjacent images, resulting in a vertical mismatch between the images along the stitched edges, producing results such as those shown in Figure 5. This figure shows the absolute surface height for adjacent images taken for the ICCWR. The software corrects for this effect by using the average height difference between edges of adjacent images to match the reference surfaces, see Figure 6. The stitching routine accounts for variable overlap between adjacent images by matching the height of the reference surface at the edges of the individual LCM images. In tests with 3013 data the matching of height along the edges of stitched images results in very little apparent error. An array of 300 stitched images is shown in Figure 7. Here, the ridges along the top-right and bottom-left borders of the right-hand image representing height relative to the local surface result from the abrupt drop-off in height as the scan passes beyond the cut edges of the sample, which is too abrupt to be captured by the polynomial reference surface.

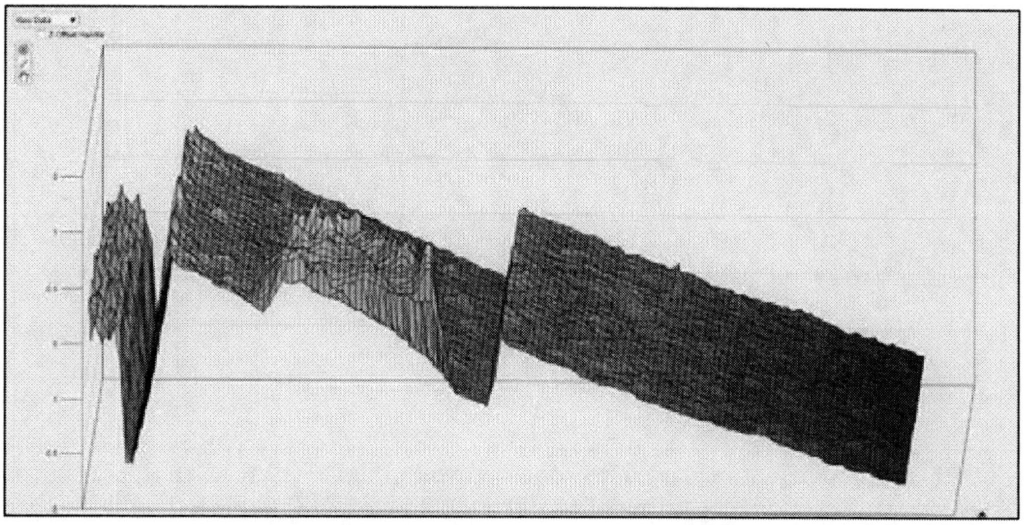

Figure 5 Vertical height mismatch between stitched images from uncorrected LCM data. Absolute surface height is shown. The x-y (area) scale is 694.92nm/unit, while the z (height) scale is 0.1nm/unit.

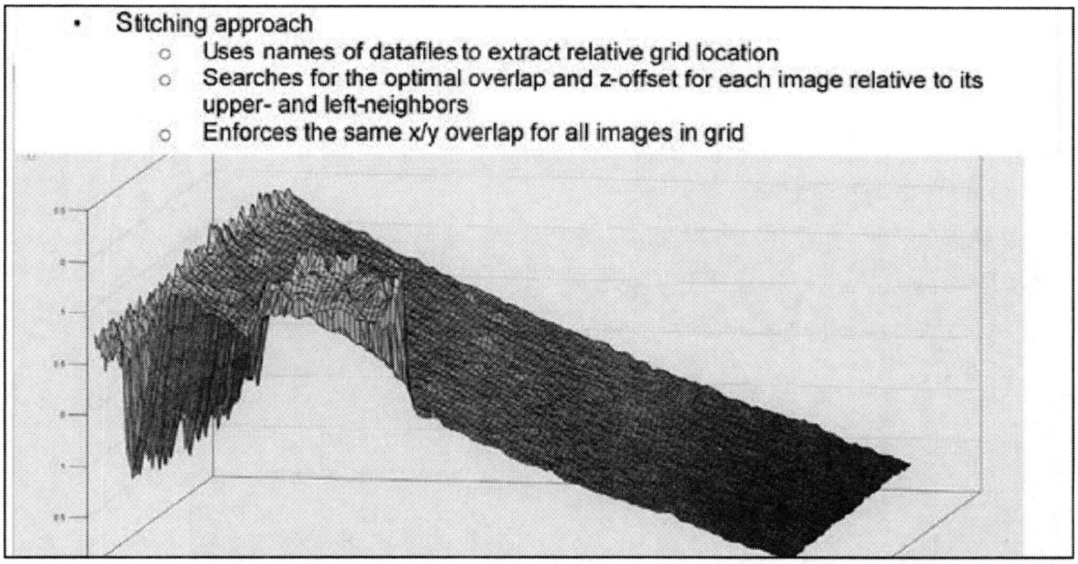

Figure 6 Correction of vertical height mismatch between stitched images using averaged height difference between edges of adjacent images. Absolute surface height is shown. The x-y (area) scale is 694.92nm/unit, while the z (height) scale is 0.1nm/unit.

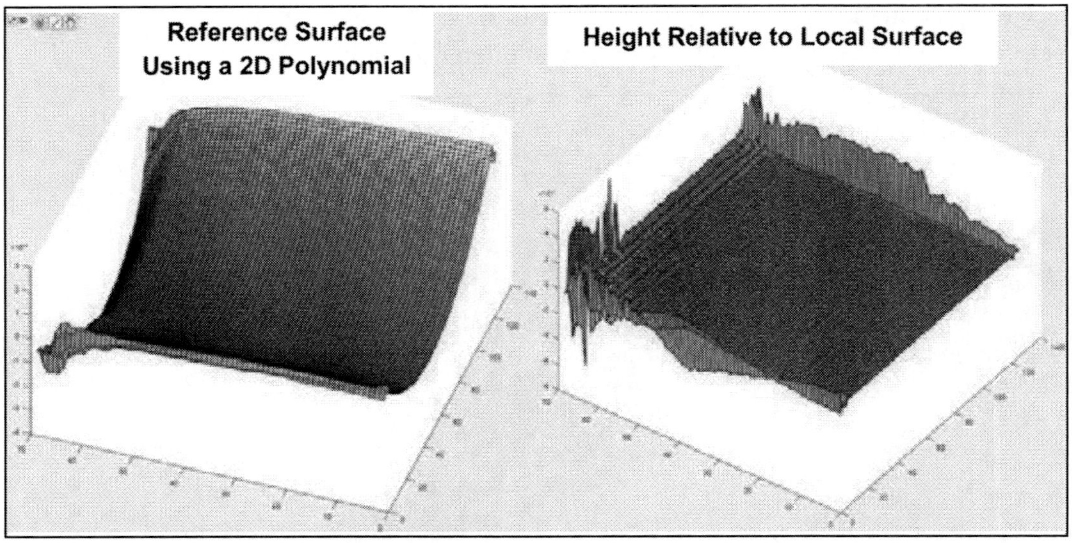

Figure 7 Array of 300 stitched height images taken for a segment of the ICCWR. The figure shows a 12.8mm x 6.9mm section, consisting of 18390 x 9980 pixels (183 Mpixels), with an x and y scale of 694.92nm for each side of a pixel. Note, without overlap the image would be 18432 x 9984 pixels.

For RGB data, assuming the same overlap as for the height data resulted in an aberration that spanned approximately 1-2 pixels. This suggests that it may be incorrect to assume that all LCM channels produce the same image-wise overlap. Possible explanations include an inherent spatial mismatch between different detectors or a distortion caused by taking LCM data through a viewport that is used to isolate radioactive canister sections. Even so, as can be seen in Figure 8, the stitching of optical images produces a mosaic that is continuous in appearance.

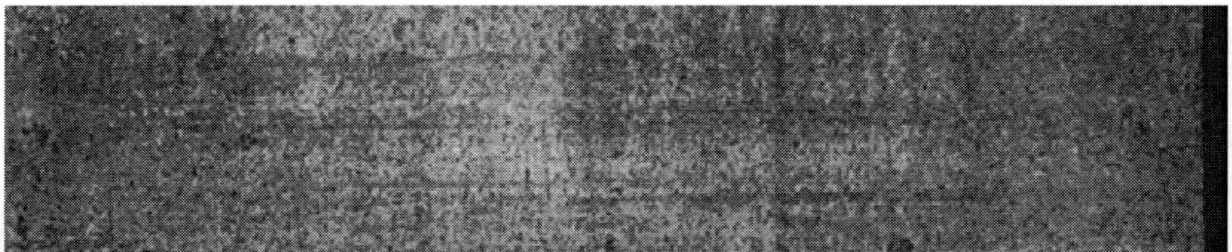

Figure 8 Array of 300 stitched optical images taken for a segment of the ICCWR. As for Figure 6, this figure represents an 12.8mm x 6.9mm section, consisting of 18390 x 9980 pixels (183 Mpixels), with an x and y scale of 694.92nm for each side of a pixel. Note, without overlap the image would be 18432 x 9984 pixels.

Results

In the current version of the image software, features are extracted using threshold-based rules that are applied after correction for sample tilt and curvature. Pits or protrusions of potential interest are identified by limits input by the user. Adjacent pixels that exceed the threshold are grouped into a single feature, which defines a region. Features flagged by the thresholds are compiled into a list for labeling and future processing. In addition to image plots and data values, the software can generate histograms of feature data, which will be important for statistical analysis in future stages of this work.

While pits and surface protrusions can be identified via height data, cracks, especially very narrow ones, require optical data for their detection. Optical features defining a crack are defined by edges. For this reason, edge detection algorithms were explored for application to RGB/optical features. One method for detection of edges in optical features is to look for large gradients in color, intensity or other picture functions. Alternative approaches include covariances of picture functions and filters. Methods considered for detection include differencing masks (Prewitt, Sobel and Roberts) and filters (Gaussian and LOG-Laplacian of Gaussian). Along with potential as edge detection methods, filters, which can be applied as masks, have the benefit of reducing image noise as well. Under certain conditions, it was found that covariances with RGB data were a bit better at edge detection than gradient techniques. However, the application of covariances relied on proper setting of thresholds. Overall, the superiority of either method was not definitively established.

Results from height and RGB covariance feature detection methods are shown in Figure 9. Figure 10 shows the difference in the size of a feature based on different LCM data. In this figure, the feature appears to have reduced area when expressed in terms of height data relative to that for optical data. The reason for this difference is that a sufficient depth is required to produce a distinct pattern in the plot of height data, while optical data relies on surface characteristics, such as roughness and color. In other words, the feature under consideration has shallow surface damage surrounding it that can be detected optically due to a difference in reflectivity (which may be wavelength dependent). The resulting optical image thus consists of the deep parts of the feature together with the surrounding shallow surface damage. However, the surface damage bounding the feature is sufficiently shallow that is does not produce much of a change in the color palette corresponding to height. Thus, the feature appears smaller when viewed in terms of height rather than optical data. The sensitivity of apparent feature size to the type of data has important implications for automated interrogation of the LCM images.

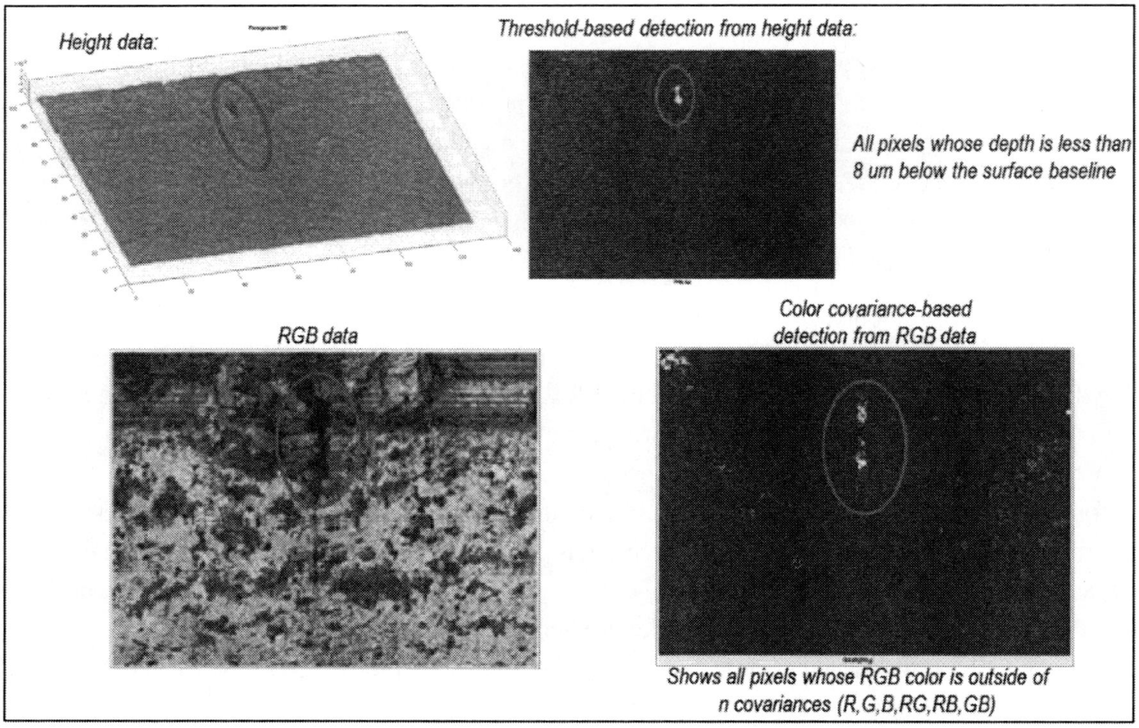

Figure 9 Feature detection via height threshold and RGB covariance threshold, shown in the upper and lower right figures, respectively. The upper left figure is local height relative to the reference surface. The lower left figure is RGB optical data.

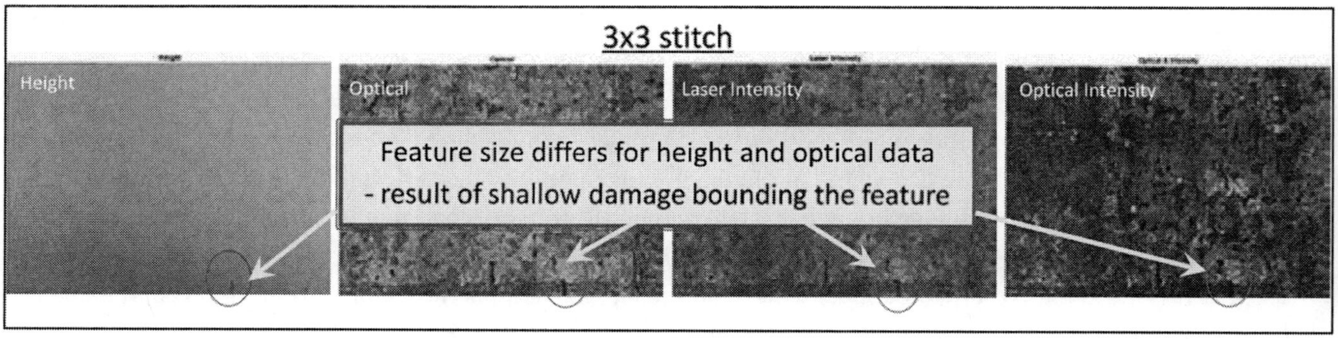

Figure 10 Difference in apparent size of a feature relative to height vs optical data. Each figure is a 3x3 mosaic.

Machine Learning Implementation

The software discussed thus far computes properties of physical features and permits user defined flagging based on specified threshold values. The next stage is the implementation and training of supervised machine learning algorithms for corrosion detection by using the output of the software along with labeled images of corrosion. The methodology will follow the path shown in Figure 11.

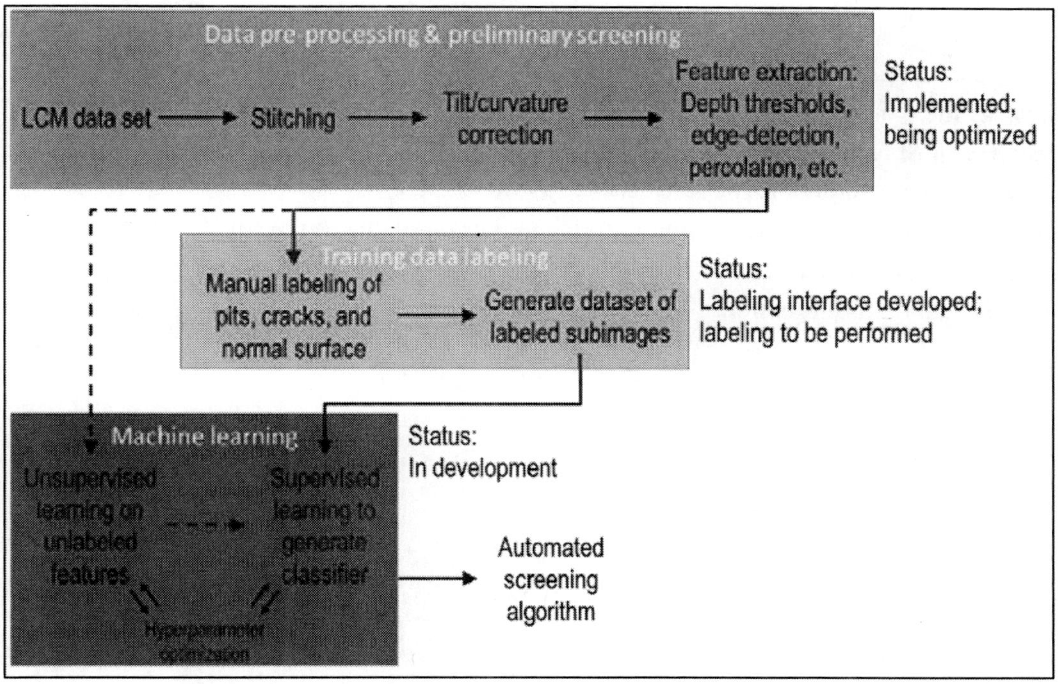

Figure 11 Methodology for application of machine learning for 3013 corrosion data.

Conclusions

This paper describes the methods used to prepare 3013 image data for application of machine learning algorithms capable of identifying corrosion events. A GUI, based on Matlab® software, was used to recover height and optical features from binary LCM output files for corrosion imaging. The software processed the sets of LCM features to compute areas, depths and shapes of surface structures, along with areal density of the structures and their associated color patterns. The structures were classified as pits, depressions, cracks and RGB functions. User specified thresholds were used to flag structures of interest, indicative of corrosion. Flagging suspect structures will significantly reduce the amount of time required for manual inspection.

As seen from Figure 10, the ability to detect and gauge surface structures is dependent on which feature data is employed. Likewise, it was found that pits could be readily identified using height data, but optical RGB data was necessary to identify cracks. Searching for cracks relied on methods used to define edges in optical images. To find the most suitable approach for identifying cracks, a number of edge detection methods were tested, including: differencing masks, filters and covariance thresholds. While covariance thresholds appeared superior in some cases, they relied on guessing the correct threshold values. If the thresholds are not properly set, either the edge would be missed, or many non-edge structures become flagged.

The software discussed in this paper is a prelude to the application of machine learning algorithms. Complex features, constructed from the LCM data, are supplied as input to the machine learning algorithms. The platforms for machine learning will be Tensorflow and Caffe. However, it is recognized that application of Tensorflow may require implementation of a Matlab-Python interface. Training sets for supervised learning will be drawn from labeled 3013 images and from coupon data.

References

1 Matlab®, Mathworks Inc.

2 Managed by Department of Nanometrology, Czech Metrology Institute. Gwyddion. [Online] [Cited: February 20, 2019.] http://gwyddion.net/.

3 Garcia, Damien. Mathworks File Exchange. [Online] [Cited: February 20, 20019.] https://www.mathworks.com/matlabcentral/fileexchange/25634-smoothn?focused=8189808&tab=function.

551422 Theoretical Modeling of Mixing Quality and Axial Hold-up Distribution for Liquid-Liquid System in Agitated Flow Mixers

New theoretical based models to describe the mixing quality and the local dispersed phase hold-up distributions in the axial direction of a continuous flow mixer have been developed and applied to predict the experimental results. The models relate the mixing index and axial hold-up distributions to a Peclet Number, the average dispersed phase hold-up, the input flow rates and the physical properties of the dispersion. The developed models offer a new understanding of the nature of the mixing process in continuous mixers with two liquid-liquid immiscible phases in flow. Experimental work was performed to measure the mixing index and to measure axial dispersed phase hold-up distribution using Laser Transmission Technique (LTT) in a mixer setup. The experimental results were used in validating the models. The experimental and the predicted data agreed well for a wide range of impeller speeds/flow ratios. Theoretical formulae to describe the local hold-up distributions in the axial direction for only one of the two phases in batch and in flow operation processes have also been developed.

Paper presented at AIChE Spring Meeting,
Final Paper #550370
March 31 – April 4 2019, New Orleans, USA Ver.1, 28/3/19

DEPLOYMENT OF DIGITAL MODELLING FOR CRACKING FURNACE ONLINE YIELD OPTIMIZATION

Yasser Al-Goufily[1], Abdullah Alqahtani[2], Abduljelil Iliyas[3], and Steve Hall[4]

(1) SABIC Technology Centre, Al-Jubail, Saudi Arabia,

(2) SABIC Technology Centre, Saudi Basic Industries Corporation, Riyadh, Saudi Arabia,

(3) SABIC Technology Centre, Saudi Basic Industries Corporation, Houston, USA

(4) Process Systems Enterprise Ltd., London, United Kingdom.

Contact: Yasser Al-Goufily, Tel: + 966 13 359 9166, Email: 29574@sabic.com

1. Introduction

Olefins plants are at the heart of the petrochemicals value chain. They convert gas and liquid feedstocks such as ethane and naphtha to ethylene, propylene and butylene. These products are then further processed into a wide range of valuable performance chemicals such as polymers, resins, esters, glycols and solvents. This conversion is performed at high temperature in thermal cracking furnaces. During the cracking process, coke is deposited in the furnace coils and transfer line exchanger.

An olefins plant is the single most energy-consuming process in the chemical industry. Modern plants have capacities up to 1.5 million tonnes/year ethylene and at this scale, small improvements in efficiency can yield significant annual savings. Olefin plant operation is thus a prime candidate for optimisation aimed at maximising economic performance.

There are several challenges that operators face regarding cracking furnaces (a) setting the desired yield for each furnace based on overall optimality, (b) understanding the extent of coking in the furnace which has a direct effect on residence time and tube metal temperatures, (c) accurately predicting ethylene yields - a key measure of plant performance and an important variable for severity control.

Another over-riding challenge is to implement improvements as part of a digital strategy, where the information contained within plant data is fully exploited by a model and the calculated results can be shared with other applications.

This paper describes how SABIC has implemented such a digitally-integrated system. The model calculates the key cracking furnace variables including conversion, product yields and degree of coking. The coking calculation uses historical plant data and a state-estimation technique to provide significantly better estimates for plant performance indicators than previous approaches. The data and calculated results are also available for other applications as needed.

2. Challenges to Operations

There are several challenges that operators face regarding cracking furnaces.

2.1 Setting the desired yield

A major challenge for operators is to determine what is the target yield in each furnace, given the fact that feed flowrates may be varying, as may the feed compositions. Also, as the proportion of ethane, propane and naphtha in the feeds varies, the number of furnaces dedicated to each feedstock type may also have to change to maintain overall economic optimality. Finally, the price of each feedstock together with prices of products and utilities may also vary. Together, all these will affect the optimal conversion to be set for each furnace.

2.2 Understanding extent and effects of coking

Coking starts from the first day of coil operation and continues until the effects are serious enough to warrant taking the unit out of service for de-coking. The degree of coking affects virtually all aspects of operation and an accurate, up-to-date prediction of coking is a critical pre-requisite for optimisation and yield accounting. The rate of coking can be exacerbated by poor operation. The effects of coke build-up increase over time and include reduced heat transfer leading to higher metal skin temperatures, increasing pressure drop and, in general, lower efficiency of operation with more energy required per unit of production. This leads to loss of yield and shortened run length. More serious is during upset operation, for example power failure, where there is significantly higher potential to damage the cracking tubes with resulting lost production and replacement costs that can run to tens of millions of dollars. Such damage can also impact safety.

2.3 Determining accurate yield prediction

The lack of accurate and timely yield information can have a significant impact on the ability to control furnace operation, leading to non-optimal operation. Given the immense cash flows in olefins plant operations, any non-optimal operation can represent a substantial missed revenue or cost-saving opportunity. An accurate measure of conversion is critical for effective furnace severity control. However, measuring the conversion and product yields from furnaces is a constant challenge because physical instruments which use extractive sampling techniques such as process gas chromatographs for composition measurement often fail or are subject to fouling. This can result from tars and contaminants created in the furnaces which can contaminate the sample lines and analyser. These affect sampling rates and sample conditioning and as a consequence effluent analysers are frequently out of order or poorly calibrated. In most cases it is not easy to maintain analysers during normal operation. In addition, each furnace will likely have its own analyser and there may, in the case of ethane/propane mixed feed furnaces, be feed analysers too. All these analysers provide many potential sources of error.

3. Simulation-based modelling approaches

There are several existing simulation-based approaches that attempt to *predict* rather than *measure* yield in real-time. Typically, yield models might use feed flow and composition, coil inlet pressures and temperatures. Often there is a limited attempt to take into account the degree of coking as it occurs throughout the production run and because coking affects conversion, the results regularly have to be 'interpreted' by experienced operators. In addition, uncertainty in any of the measured variables from the plant is not taken into account, mainly because there is no way using conventional approaches to incorporate such uncertainty into any predictive calculations. Hence there is a risk of inaccurate values being calculated for critical parameters such as yield, pressure drop and maximum tube metal temperature,

Sometimes yield models are used in conjunction with coking models, with coking over the run length simulated by a semi-empirical coking model using presumptive coking kinetic parameters fixed beforehand. The available models do not generally predict coking accurately, especially given that coke build up varies radially, axially and over time. Coke predictions tend to depart from reality over the normal course of a run, particularly if there is deviation from standard operation. Consequently such models lack the ability to predict yield and tube metal temperature to a sufficient degree of accuracy.

In some cases, soft sensors based on statistical models[1] are used. Again, these reflect 'normal' operation but have no predictive capability, so can deviate from reality if there is some type of upset in the operation.

4. Advanced process modelling approach for accurate yield prediction and coking monitoring

The digital model-based approach described in this paper is based on a novel Advanced Process Modelling (APM) methodology[2] that couples detailed predictive furnace models with overall olefins plant models and regularly-updated plant data.

4.1 Determining desired furnace yields

The following figure shows the scope of the overall olefins plant model. It covers not only cracking furnaces, but also quench, compression and downstream separation. All these have an impact on operating costs.

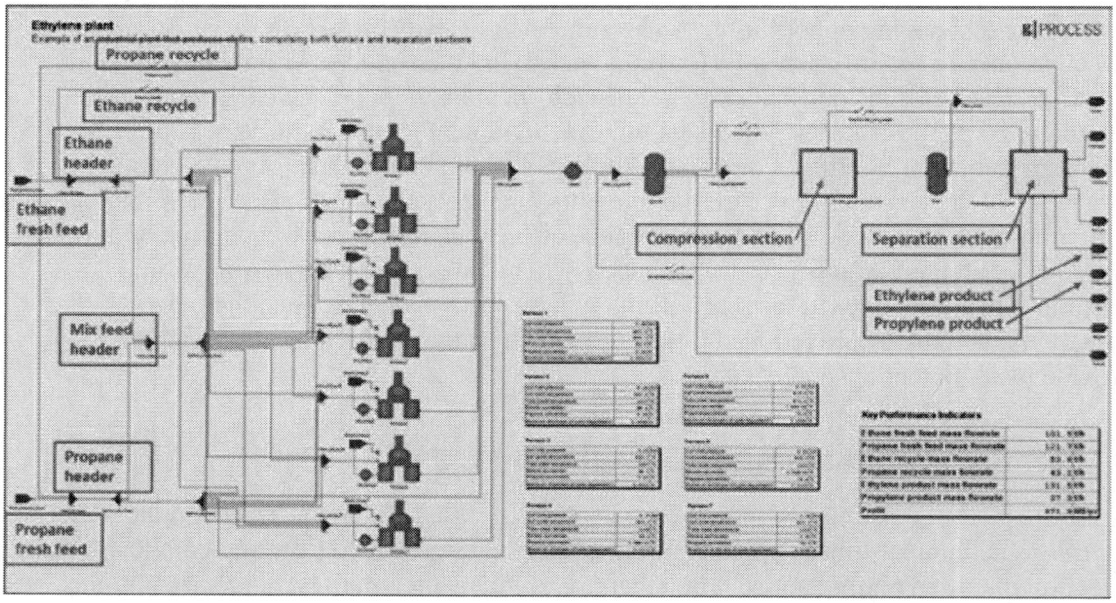

Figure 1 – Overall olefins plant model

---

[1] Soft sensors for processing plants, by G. D. Gonzalez, published in the Proceedings of the Second International Conference on Intelligent Processing and Manufacturing of Materials, July 10-15, 1999, Honolulu, Hawaii

[2] First-principles modelling for optimization of process operations, Costas Pantelides, Foundations of Computer Aided Process Operations / Chemical Process Control, Tucson, Arizona, January 8-12, 2017

The predictive cracking furnace model includes first-principles representation of coil geometry, material and heat balances, cracking kinetics and coking kinetics. The plant data used by the model are measured variables such as feed rate, feed composition, coil inlet temperature and coil outlet temperature, together with any additional available measurements that can be used for real-time calibration of the model, such as pressure drop over coils or tube metal temperatures. Uncertainty is assigned to these additional measured variables and these are considered in the overall reconciliation of the plant data.

In addition to feed flows and compositions, feed, product and energy pricing is considered in the overall economic model. The whole plant model runs and determines the desired yield required in each furnace. The model can be run regularly or whenever feedstocks or pricing change.

4.2 Determining the degree of coking

The furnace system models are solved using dynamic state estimation techniques based on an Extended Kalman Filter approach to provide up-to-date estimates of process state variables, selected process parameters and other key performance indicators (KPIs). The use of Kalman filtering as feedback in process control is not new. However, the novelty here comes from combining the use of Kalman filters with rigorous process simulation models.

The approach exploits data redundancy features within the data system and the applied confidence limits to measured variables to reduce overall model uncertainty, i.e. the approach makes maximum use of all the data available to it. This powerful technique allows the model to be used, if necessary, beyond its initial operating window. Kinetic parameters of the coking model are adjusted periodically to match observed reality as new evidence arrives in the form of actual plant measurements such as pressure drop over the cracking coils. Effectively this results in a *self-calibrating coking model* that can accurately predict the current state of coking along the length of the coil and take this into account in the yield prediction calculation.

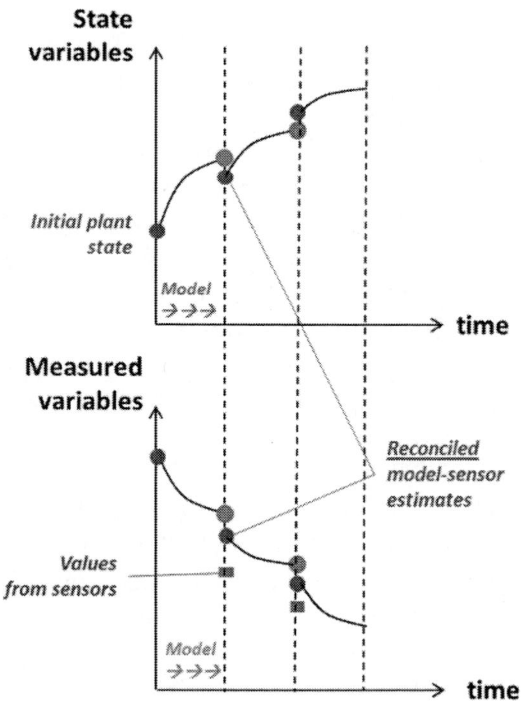

Figure 2 – State estimator operation

The operation of a state estimator in conjunction with a process model is illustrated in Figure 2. At the beginning of the first interval, the plant model has what it believes are correct values for the plant state (e.g. pressures, state of coking) and various measurements. It then calculates new values for the next time interval (orange dots) based on this understanding. There is an element of error in the new prediction, because no model is ever perfect.

In the meantime, new measurements (grey rectangles) arrive from the plant sensors. These will typically differ from the values calculated by the model; they also carry a degree of uncertainty, reflecting, for example, instrument error.

The state estimator then uses mathematical and statistical techniques to find the best 'reconciled' values for all the variables that it is allowed to adjust (blue dots). This effectively simultaneously takes into account uncertainty in plant measurements as well as possible errors in the model representation (for example, resulting from an inaccurate coking model), to provide a reconciled plant model. This process continues for each subsequent time-interval.

At the end of each interval, the model contains the best up-to-date reconciled information that can be achieved for the process, subject to the accuracy of the model representation and the uncertainty in the measurements.

4.3 Determining furnace yields

In order to be deployed effectively by the state estimator, the underlying furnace model requires a certain minimum level of fidelity to calculate accurate yields. In particular it is important that it has a high degree of *predictive* capability. The model needs to include, at a minimum:

- **Furnace geometry.** The coil geometry needs to be represented accurately, along with its material properties such as thermal conductivity.
- **Heat input to cracking coils.** It was found in offline studies that using a representative relative heat release profile according to the furnace design together with a severity specification provides sufficient fidelity on modelling the firebox side of the furnace when the focus is on prediction of cracking yields and coking rates along the tubes. Imposing a relative heat flux profile can therefore be used as a much simpler and faster approach that is very effective for representing the actual heat input, as opposed to using more rigorous methods such as zonal methods or computational fluid dynamic-like approaches.
- **Detailed cracking kinetics**. Accurate kinetics for the cracking reactions are critical to reliable modelling of the furnace and indeed the whole ethylene plant performance. There are many ways of supplying kinetics: kinetic calculations supplied with the modelling package; commercially-available kinetics packages; or via users' own kinetics, perhaps fitted to plant or laboratory data. In the modelling tool used, supplied kinetics based on literature sources can be adjusted to match observed plant or pilot plant data.
- **Coking model**. Again, the coking model is important as coking starts from day one of a production run and affects virtually all aspects of operation and economics. Given the many variables in furnace operation, it is virtually impossible to predict kinetics accurately from a static model; this is a key reason for the approach described here. Well-established kinetic models from literature were used, with parameters such as rate constants periodically updated based on observed plant performance, using dynamic state estimation techniques.

A furnace model fulfilling these requirements was implemented in SABIC under the name 'gCRACKER' (also referred to by PSE as gPROMS Olefins Cracking Monitor). Figure 3 shows some of the furnace configuration screens related to coil selection and definition of the coking kinetic model.

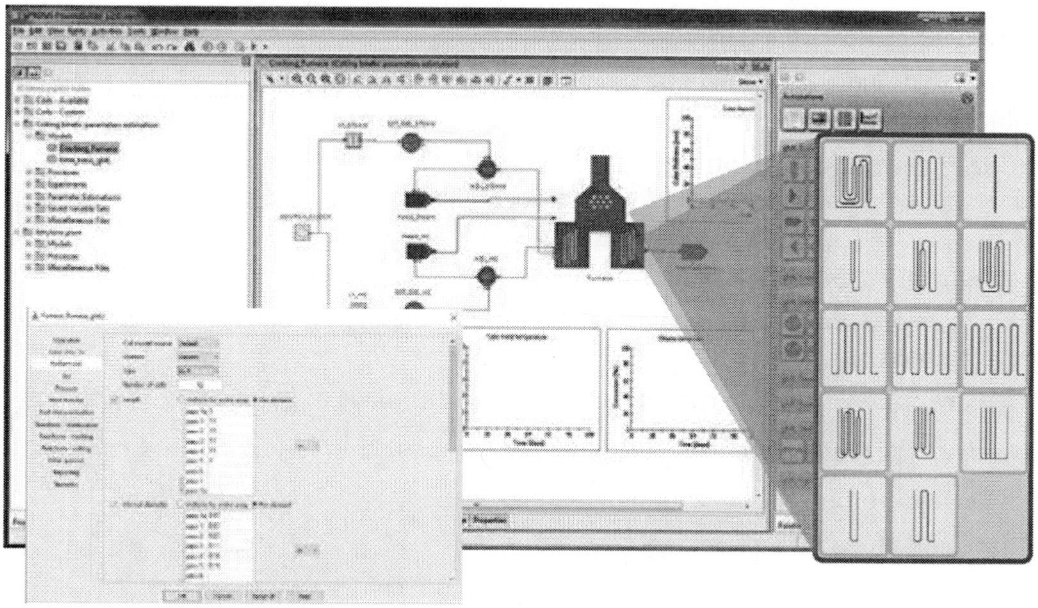

Figure 3 – Ethylene furnace setup showing coking kinetics definition dialogs

5. Experience in SABIC crackers

The technology has been applied to a large ethylene plant in Saudi Arabia processing mainly ethane and propane feedstocks.

5.1 Workflow

In implementing the gCRACKER model in SABIC's plants, a number of key elements are included in the workflow, as represented in the following figure.

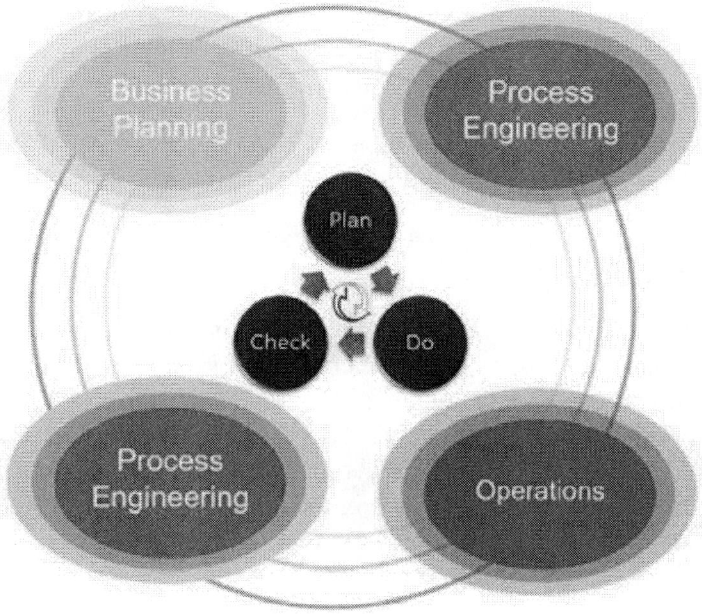

Figure 4 – Overall workflow to deploy a digital olefins plant model in SABIC

The various workflow elements are as follows:

- Business planning – to provide a weekly run plan considering the required production volumes and costs
- Process engineering - to utilize the gCRACKER optimization model and determine the operating parameters which maximize profitability
- Operations – to run the furnaces with gCRACKER at the optimized conversion and highlight any operating limitations
- Process engineering – to estimate the benefits considering the contribution margins of the generated products

These various workflow elements ensure that the best use is made of the model.

The following optimization parameters were used in the model:

- Ethylene and propylene contribution margins
- Ethane and propane conversions
- Ethane and propane recycle flows

5.2 Results

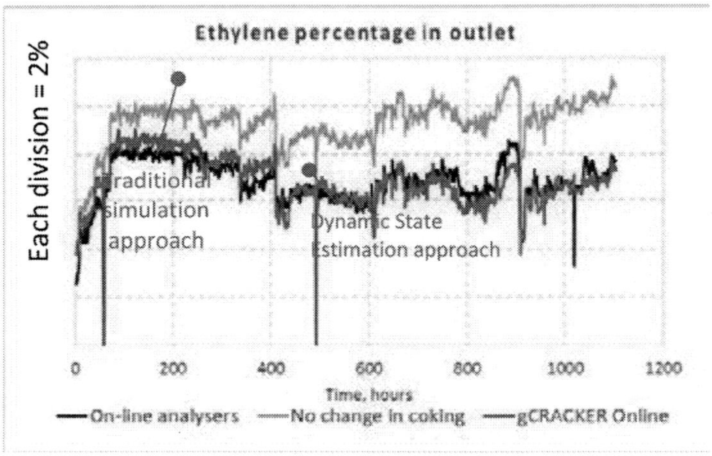

Figure 5 – Ethylene prediction (red and grey line) compared with plant data (black line)

The plot in Figure 5 shows results from both gCRACKER (CrackingMonitor) and a simulation case, compared to plant data. The data were taken from a period of operation where reasonably reliable ethylene yield measurements were available and could therefore be used to provide verification of model predictions.

During this performance run, great care was taken to ensure the relevant analysers were working well, in a calibrated state with a working sampling system. Two sets of results are compared against the on-line analyser data (black line) over a 50-day run.

The grey line shows the ethylene yield when the coking is simulated assuming fixed coking parameters and no change to coking rate. The conditions in the furnace are simulated as the run progresses, taking into account actual feed flowrate, controls, etc., to determine coking and ethylene yield. This approach, which is used by current commercial software tools, works well in the early stage of the run. However, the inability of the coking model to accurately reflect all events affecting coke deposition over the course of the run eventually causes the calculated coking state and hence ethylene yield prediction, to deviate significantly from reality.

By contrast, the red line shows the prediction using the dynamic state estimation techniques. This is more accurate than simply simulating the operation, as the plant data are used to adjust the prediction of coking progressively as the run unfolds – i.e. the coking model is effectively *self-tuning*, meaning that the degree of coking reported is always accurate and up-to-date.

Note that the ethylene yield measurement recorded by the on-line process analysers was *not* included in the state estimation approach so the ethylene concentration is a purely *predicted* value. It can be seen that the predictions reflect the actual measured values very closely.

5.3 Benefit realization

Having determined the optimal conversion required in each furnace through the overall olefins plant optimizer, the closed-loop on-line Gcracker (CrackingMonitor) tool was employed. The optimization parameters were set and the conversion set-points relayed to the APC controllers.

The realized benefit of online optimization was an increase in olefins production by **2.8** tons / hour which translates to a yield enhancement of **2%**.

Ethylene Benefit (tph)	+ 0.3
Propylene Benefit (tph)	+ 2.5
Olefins Yield Increase %	**+ 2%**

Table 1. Benefits of implementing the gCRACKER (CrackingMonitor) tool

6. Conclusions

With the whole olefins plant model and optimizer, linked to furnace model and state estimator, all connected to the plant's data acquisition system, plant data is processed in real time. This digital approach provides:

- highly accurate predictions of conversion and product yields, which can be used for furnace severity control
- accurate estimates of the current degree of coke deposition in each section of the furnace coil, for use in optimising de-coking intervals
- accurate soft-sensed measurements of furnace KPIs or any other variables of interest in the plant that can be used for reporting or for improved control
- an up-to date model of each furnace that, when included with a high-fidelity model of the surrounding process incorporating updated plant parameters, can be used for whole-plant optimisation (online or offline), look-ahead simulation and many other activities.

In terms of benefits, this results in:

- significantly improved confidence in the reliability of furnace effluent composition values

- the ability to accurately control the furnace based on conversion as opposed to, for example, Coil Outlet Temperature
- improved furnace control, leading in turn to higher throughput and profitability
- better control and management of coking, leading to higher efficiency and longer run lengths
- early warning of excessive coking, helping to avoid split tubes and other unplanned outages which can be very costly
- better, up-to-date process models on which to base optimisation, leading to better, more profitable operation
- accurate soft-sensed values for all variables within the scope of the process model, leading to greater understanding and monitoring of operation
- improved consistency between off-line and on-line models which aid in accurate predictions for production planning and control.

A key further application of the approach outlined here is in providing an up-to-date plant model – including accurate coking prediction and adjusted reaction kinetics – for a real-time plant optimiser. This is a crucial requirement for extracting the maximum value from online optimisation calculations.

The realized benefit of online optimization was an increase in olefins production by **2.8** tons / hour which translates to a yield enhancement of **2%**.

551016 Optimization of Mixed Smr Plant Using Advanced Modelling & Six Sigma Tools to Enable Smart Manufacturing: Case Study

Synthesis gas is the key feedstock for many industrial applications especially for ammonia, methanol, Oxo-alcohols etc. Most of the synthesis gas produced in the world is from natural gas using steam methane reforming (SMR). Mixed SMRs have carbon di-oxide added to the feed gas to produce synthesis gas with lower H2/CO ratio than that of conventional synthesis gas applications. As the gas reserves are depleting in the world, existing reforming plants are open to use natural gas from different sources. Different natural gas sources offer different compositions in the gas, which significantly affect the performance of the mixed steam methane reformers (SMR) in terms of reforming efficiency, energy consumption, H2/CO ratio in synthesis gas etc. Effect of feed gas disturbances in mixed SMRs is significant as these plants are operated at higher reforming temperature and lower steam to carbon ratios. In order to operate the reforming plant efficiently, optimally and smartly for all feed gas disturbances, optimization of the plant using advanced modelling & six sigma tools, precise pro-active control, continuous operations monitoring are necessary.

In this paper, an implemented optimization of mixed steam reformer in the oxo-alcohol production process case study is presented. An Aspen Plus model based study has been performed to optimize the mixed SMR plant operations for the existing feed gas composition. Then, coupled the model-based study with six sigma tools to develop the best operating conditions and zone for various scenarios. A number of contour diagrams have been developed for these scenarios using MINITAB, and response surface optimizer has been developed using response surface methodology. In this part, optimized the steam to carbon ratio, carbon dioxde to natural gas ratio, using surface response methodology using MINITAB, and the optimized conditions are translated into operator language and implemented in the systematic procedure. The implementation of the optimized conditions resulted in savings of energy 0.22 MMBTU/Ton of product and production increase of 2000 TPA.

Furthermore, a detailed dynamic model, as in the DCS of the plant, has been developed ASPEN DYNAMICS and analyzed the effects of the feed gas disturbances on the process. The analysis showed that H2/CO ratio in synthesis gas, reformer energy efficiency, methane slippage, and oxygen content in the stack gases would have serious repercussions and dynamic profiles have been developed for various feed gas compositions changes. In addition, it has been understood that a proactive advanced feed forward controls are necessary for reformer temperature, methane slippage, H2/CO ratio along with the existing feedback controls. Feed forward based control loops have been added on top of the existing controls based on the carbon number and calorific value of the feed gas. These upgradations enable the smooth operation of the plant during drastic feed gas composition changes. The added advanced process control would eliminate the situation of process upset and un-planned shutdowns, that result in improved equipment integrity and reliability.

553432 Molecular Analysis Reveals the Myth of Value Chain and Maximize Potential

Process synthesis is about finding the appropriate processing routes with selection of catalyst, reaction and separation technologies followed by flowsheeting (connection of selected technologies) and equipment design at the least capital and operating cost. The common feature of current process synthesis methods is that they are mainly based on macroscopic parameters but do not exploit microscopic effects at molecular level. As a result, the fundamental aspects cannot be fully exploited to discover novel ideas. However, the essence of process synthesis is about maximizing the production of desirable chemical species via optimizing the routing of key species to their most appropriate destinations. To achieve this goal, a molecular analysis methodology for process synthesis is proposed. Several industrial case studies are provided to prove the concept and demonstrate the value.

550027 Thermodynamic Modeling of Lithium Salts with the Electrolyte NRTL Model Using Hydration Chemistry

The lithium bromide-water solution is a popular working fluid due to its environmentally friendly, nonvolatile, and non-toxic properties; however, this solution has issues with corrosion and crystallization. One proposed solution is to modify the lithium bromide-water working solution by adding anticrystallization and anticorrosion agents such as lithium chloride, lithium iodide, and lithium nitrate. Thermodynamic models are essential to facilitate the advancement of process design, optimization, and operation range of absorption heat pumps and refrigeration units that utilize lithium salt systems. To obtain a rigorous thermodynamic model of lithium salt solutions, appropriate solution chemistry must be established. The current work proposes that the lithium ion is present with permanently bound water molecules that should be considered as part of the ion. The hydration chemistry provides an increased accuracy to the symmetric electrolyte NRTL (eNRTL) activity coefficient model. This solubility model correlates the nonideality of the solution with the composition dependency by utilizing two binary interaction parameters for each of the electrolyte-electrolyte and molecule-electrolyte pairs in the system. Each parameter is determined by applying up to three temperature dependent coefficients and then regressed from experimental data. This allows the parameters to establish the accuracy of the model for concentrations up to saturation and temperatures between 273.15-473.15 K.

550607 Multiscale Process Simulation By Combining Subgrid Modeling and Computational Fluid Dynamics

Many realworld processes are multi-scale processes, which means that they have important features at different scales in time and/or space. Typically not all features can be resolved, so that for simulations of such processes certain assumptions or simplifications have to be made and so--called subgrid models are used to capture the effects of one scale onto the other. The benefit of combining different scales was previously shown e.g. for a wet granulation process [1].

In this contribution we show a new automated coupling method between the simulation tools STARCCM+ by Siemens PLM and gPROMS by PSE, where STARCCM+ solves for all flow properties and gPROMS evaluates the fine scales based on the solution provided by STARCCM+. The coupling between both codes is done in an automatic way. This includes the domain decomposition, the calculation of and the transfer of all relevant variables.

The details of the coupling are discussed and the benefit of this approach is demonstrated based on used cases like a liquid-liquid emulsion for PVC production in a full scale slim vessel and a spray drying example. The numerical results are compared against experimental results [2] and the effect of different ways of decomposing the domain is discussed.

References:

1. Barosso, D. et al.: A multiscale, mechanistic model of a wet granulation process using a novel bidirectionalPBM–DEM coupling algorithm, Eng. Sci. 2015, 123, 500–513
2. Maass, S. et al.: Flow Field Analysis of Stirred Liquid-Liquid Systems in Slim Reactors, Chem. Eng. 2011, 34, No. 8, 1215–1227

550382 Dynamics and Control of Reactive Distillation Column with External Recycle

Reactive distillation columns with external recycle between the top and bottom (RDCs-TBER) have a better steady-state performance as compared with conventional reactive distillation columns (CRDCs) for the separations of reacting mixtures featuring the most unfavorable ranking of relative volatilities (i.e., the reactants are the lightest and heaviest components with the generated products in between). In this article, their dynamics and control are studied in great detail, with special attention given to the influences of external recycle on process dynamics and controllability. Three kinds of control systems are developed. The top and bottom levels of all control systems are controlled by reflux flow rate and reboiler heat duty, respectively. The only distinguish of these three control systems is that one is with constant external recycle, another is with constant reflux ratio, and the last is controlling the product purity with external recycle flow rate.

Two reactive distillation systems, executing a hypothetical reversible reaction, A + B ↔ C + D ($\alpha_A > \alpha_C > \alpha_D > \alpha_B$), and lactic acid esterification, are employed to inspect the dynamics and controllability of the RDC-TBER. The results show that the supplementary manipulated variable (i.e., the flow rate of the external recycle) can be used to improve the process operation. The RDC-TBER, on the other hand, shows greatly improved tracking performance with even enhanced disturbance rejection capabilities.

AUTHOR INDEX

AIChE
120 Wall Street, FL 23
New York, NY 10005-4020

ISBN 978-1-5108-8697-1
9 781510 886971

CURRAN ASSOCIATES INC.
proceedings.com